ASTRONOMY
101

FROM **THE SUN AND MOON** *TO* **WORMHOLES AND WARP DRIVE, KEY THEORIES, DISCOVERIES, AND FACTS ABOUT THE UNIVERSE**

CAROLYN COLLINS PETERSEN

Adamsmedia
Avon, Massachusetts

Published by
Adams Media, a division of F+W Media, Inc.
57 Littlefield Street, Avon, MA 02322. U.S.A.
www.adamsmedia.com

ISBN 10: 1-4405-6359-4
ISBN 13: 978-1-4405-6359-1
eISBN 10: 1-4405-6360-8
eISBN 13: 978-1-4405-6360-7

Printed in the United States of America.

10   9   8   7   6   5   4   3   2

Cover image © Jupiterimages Corporation and 123rf.com.
Interior illustrations by Claudia Wolf.

*This book is available at quantity discounts for bulk purchases.
For information, please call 1-800-289-0963.*

DEDICATION

To my many astronomy teachers. You promised me the Moon,
planets, stars, and galaxies—and the universe delivered!

ACKNOWLEDGMENTS

Thanks to my "brain trust" of astronomer friends and colleagues: Natalie Batalha, Jack Brandt, Tania Burchell, Lynn Cominsky, Dennis Crabtree, Jack Dunn, Davin Flateau, David Grinspoon, Madulika Guthakurtha, Nicole Gugliucci, Christine Pulliam, Catherine Qualtrough, Seth Shostak, Mike Snow, Padma Yanamandra-Fisher, and to the editors at Adams Media. Everyone made very cogent and helpful suggestions along the way. Special thanks go, as always, to Mark C. Petersen for reading the entire manuscript, making excellent comments, and providing love and support during the creative process.

CONTENTS

INTRODUCTION

Welcome to *Astronomy 101* and one of the most fascinating sciences in the universe! Whether you're familiar with the night sky and want to learn more about what's "out there" or just beginning your cosmic journey of understanding, there's something here to teach and inspire you.

I've been "into" astronomy since I was a small child when I would go out with my parents to see what was "up out there." I grew up wanting to become an astronaut and eventually spent time in college studying a lot of astronomy and planetary science. In the early part of my career I did astronomy research (mostly into comets), and that experience taught me there is *nothing* quite so fascinating as standing (or sitting) in front of the cosmos, open to new discoveries in space! Nowadays, I spend my time communicating astronomy to the public because I want others to experience the thrill of wonder and discovery that keeps astronomers going. I often give presentations about astronomy on cruise ships and other public places, and the questions people ask about the stars and planets are always fascinating and well thought out. It shows me that the love of the stars is bred into all of us—and makes us want to know more about the cosmos.

In *Astronomy 101*, you get a taste of the cosmos. Astronomy is the scientific study of objects in the universe and the events that shape them. It is one of the oldest sciences and dates back to a point in human history when people first looked up at the sky and began to wonder about what they saw. Astronomy tells how the universe works by looking at what it contains. The cosmos is populated with stars, planets, galaxies, and galaxy clusters, and these are all governed by measurable physical laws and forces.

Each topic in this book gives you a taste of the subject it covers, from planets out to the most distant objects in the universe, introducing you to some of the people who have done astronomy, and venturing into some "far out" topics, such as extraterrestrial life and the science-fiction universes familiar to TV viewers, moviegoers, video gamers, and readers. Throughout the book, I've woven in some basic concepts about astronomy and space, such as how orbits work and how to calculate distances in space.

Finally, although this isn't a "how to" book, in the final chapter, I leave you with a few thoughts about how to go about exploring the universe from your backyard and how we can all work together to mitigate light pollution—the scourge of all sky gazers.

You can read this book from start to finish, or pick and choose the topics you want to read. Each one gives you a unique insight into the endlessly fascinating universe. And, if what you read spurs you on to more investigation, the reference section at the back points you to further reading.

Why Do People Do Astronomy?

The astronomer Carl Sagan once said that modern people are descendants of astronomers. Humans have always been sky watchers. Our earliest ancestors connected the motions of the

Sun, Moon, and stars to the passage of time and the yearly change of seasons. Eventually, they learned to predict and chart celestial motions. They used that information to create timepieces and calendars. Accurate knowledge of the sky has always helped navigators find their way around, whether across an ocean or in deep space.

Humanity's fascination with the sky may have begun with shepherds, farmers, and navigators using the sky for daily needs, but today that interest has blossomed into a science. Professional astronomers use advanced technology and techniques to measure and chart objects and events very precisely. New discoveries come constantly, adding to a priceless treasury of knowledge about the universe and our place in it. In addition, the tools and technologies of astronomy and space exploration find their way into our technologies. If you fly in a plane, use a smartphone, have surgery, surf the Internet, shop for clothes, eat food, ride in a car, or any of the countless things you do each day, you use technology that in some way derived from astronomy and space science.

In my childhood I was enthralled with a 1927 poem written by American writer Max Ehrmann, "Desiderata." My favorite line from it is: "You are a child of the universe, no less than the trees and the stars; you have a right to be here." That's why I draw a link between space and our DNA. All living things are a direct result of the processes that created the cosmos, built the galaxies, created and destroyed stars, and formed planets. We are, in many senses of the term, star stuff. Every atom of every living thing on Earth originated in space, and it's poetic and delightful that we evolved to look back out at the light from stars that will eventually contribute their own "stuff" to create other stars, planets, and maybe even life. That's why I can say that a love of the stars is woven into our DNA. Whether we're professional astronomers or casual sky gazers, that's what draws our attention back to the depths of space. It's where we came from.

Welcome home!

ASTRONOMY LINGO

Throughout this book I use some astro-lingo, so let's look at a few definitions that will help you understand the language.

DISTANCE

Distances in astronomy get very large very fast. Astronomers use the term *astronomical unit* (shortened to *AU*) to define the distance between Earth and the Sun. It's equivalent to 149 million kilometers (93 million miles). (Astronomy is done in metric units.) So, for distances inside the solar system, we tend to use AU. For example, Jupiter (depending on where it is in its orbit) is 5.2 AU away from the Sun, which is a distance of 774.8 million kilometers (483.6 million miles).

In interstellar space, we use other units. The *light-year* (shortened to *ly*) comes from multiplying the speed of light, 300,000 kilometers per second, by the total seconds in a year. The result is the distance light travels in a year: 9.5 trillion kilometers. The nearest star is 4.2 light-years away from us. That means that it's four times 9.5 trillion kilometers, which is a huge number. It's easier to say that the star is 4.2 light-years away.

Astronomers also use the term *parsec* (or *pc* for short). One parsec equals 3.26 light-years. The famous Pleiades star cluster is around 150 parsecs (about 350–460 light-years) away. The nearest spiral galaxy, called the Andromeda Galaxy, is about 767 kiloparsecs or 2.5 million light-years from us.

Really huge distances are measured in terms of *megaparsecs* (millions of parsecs, or *Mpc*). The closest cluster of galaxies to our own Milky Way Galaxy lies about sixteen megaparsecs, or nearly 59 million light-years, away. The very largest distances are measured in units of *gigaparsecs* (billions of parsecs, *Gpc*). The limit of the visible universe lies about 14 Gpc away from us (about 45.7 billion light-years).

LIGHT

Light is the most basic property we study in astronomy. Studying the light emitted, reflected, or absorbed by an object tells a great deal about the object. The speed of light is the fastest velocity that anything can move in the universe. It is generally stated as 299,792,458 meters (186,282 miles) per second, in a vacuum. It has been measured very accurately and is the standard that astronomers and physicists use. However, as light passes through water, for example, it slows down to 229,600,000 meters (140,000 miles) per second. The letter *c* is shorthand for the speed of light.

Lightspeed does more than define distances. They help us get an idea of the age of the universe. Step out and look at the Moon. The "Moon" your eyes see is 1.28 seconds old. The Sun is 8.3 light-minutes away—you see the Sun as it was 8.3 minutes ago. The light from the next closest star, called Proxima Centauri, shows us how it looked just over four years ago. The light from a galaxy that lies 65 million light-years away left that galaxy when the dinosaurs were facing extinction. The most distant objects and events existed when the universe itself was only a few hundred thousand years old. Astronomers see them as they *were* more than 13 billion years ago. When you look out in space, you're looking back into time. The farther across space you look, the further back in time you see. This means the telescopes and instruments we use to study the cosmos are really time machines.

Redshift and Blueshift

In several places in the book, you can read about spectra and spectroscopy. These are important tools in the astronomer's kit. They take light from an object and slice it up into its component wavelengths. The result is a spectrum. If you've ever passed light through a prism and noticed the spread of colors that come out, you've seen the principle that spectroscopy works under—except, there are *more* "colors" in the spectrum that we can't see. Our eyes can only detect the colors you see coming from the prism.

Visible Light

Visible light consists of the following colors:

- Red
- Orange
- Violet
- Yellow
- Green
- Blue
- Indigo

Spectroscopy gives information about how fast or slow an object is moving, and helps astronomers figure out how far away it is. An object moving toward us shows lines in its spectrum that are *blueshifted*, or shifted to the blue end of the spectrum. If it moves away from us, then the lines are *redshifted*—shifted toward the red end of the spectrum. The term *redshift* is often used to indicate an object's distance, say a redshift of 0.5. Astronomers notate that as $z = 0.5$.

THE SOLAR SYSTEM

Our Neighborhood in Space

The solar system is our local place in space. It contains the Sun, eight planets, several dwarf planets, comets, moons, and asteroids. The Sun contains 99.8 percent of all the mass in the solar system. In orbit around it, in a region called the *inner solar system*, are:

- Mercury
- Venus
- Earth
- Mars

Beyond Mars, there's the Asteroid Belt, a collection of rocky objects of various sizes. The *outer solar system* is dominated by four giant planets that orbit the Sun at large distances and contain about 99 percent of the *rest* of the solar system's mass. These are:

- Jupiter
- Saturn
- Uranus
- Neptune

In recent years, planetary scientists (people who study solar system bodies) have focused much attention on a region beyond the gas giants called the *Kuiper Belt*. It extends from the orbit of Neptune out to a distance of well beyond 50 AU from the Sun. Think of it as a very distant and much more extensive version of the Asteroid Belt. It's populated with dwarf planets—Pluto, Haumea, Makemake, and Eris, for example—as well as many other smaller icy worlds.

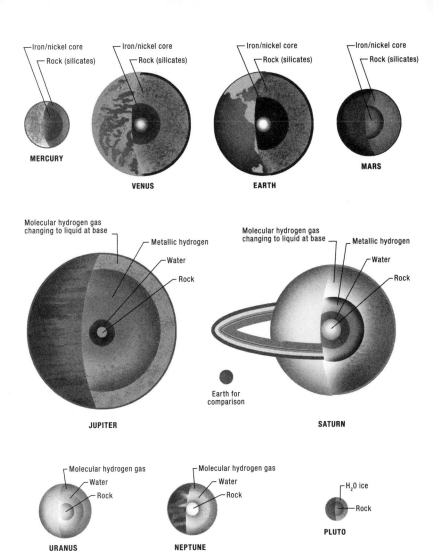

The general compositions of the worlds of our solar system.

Inventory of the Solar System

1. 1 star
2. 8 main planets
3. 10 (and counting) dwarf planets
4. 146 (and counting) moons
5. 4 ring systems
6. Countless comets
7. Hundreds of thousands of asteroids

Planetary scientists often refer to the inner worlds of the solar system as the "terrestrial" planets, from the word *terra*, which is Latin for "earth." It indicates worlds that have a similar rocky composition to Earth. Earth, Venus, and Mars have substantial atmospheres, while Mercury appears to have a very thin one. The big outer worlds are called the "gas giants." These planets consist mostly of very small rocky cores buried deep within massive spheres made of liquid metallic hydrogen, and some helium, covered by cloudy atmospheres. The two outermost planets—Uranus and Neptune—are sometimes described as "ice giants" because they also contain significant amounts of supercold forms of oxygen, carbon, nitrogen, sulfur, and possibly even some water.

The Oort Cloud

The entire solar system is surrounded by a shell of frozen bits of ice and rock called the Oort Cloud. It stretches out to about a quarter of the way to the nearest star. Both the Kuiper Belt and the Oort Cloud are the origin of most of the comets we see.

Moons and Rings

Nearly all the major planets, some of the dwarf planets, and some asteroids have natural satellites called *moons*. The one we're most familiar with is Earth's Moon. The lunar (from the name *Luna*) surface is the only other world on which humans have set foot. Mars has two moons, called Phobos and Deimos, and Mercury and Venus have none.

The gas giants swarm with moons. Jupiter's four largest are Io, Europa, Ganymede, and Callisto, and they're often referred to as the Galileans, honoring their discoverer, astronomer Galileo Galilei. Over the past few decades, at least sixty more have been discovered orbiting Jupiter. Saturn, Uranus, and Neptune also sport dozens of smaller icy worlds. Out in the Kuiper Belt, dwarf planet Pluto has at least five satellites, while Eris has at least one.

Each gas giant planet has a set of rings. Saturn's is the most extensive and beautiful. It's possible Earth had a ring in its early history, and planetary scientists now look at rings as somewhat ephemeral (short-lived) objects.

Orbiting the Sun

All the planets of the solar system travel around the Sun following paths called orbits. Those paths are defined by Kepler's Laws, which state that orbits are elliptical (slightly flattened circles), with the Sun at one focus of the ellipse. The farther out an object orbits, the longer it takes to go around the Sun. If you connect a line between the Sun and the object, Kepler's Law states that this line sweeps out equal areas in equal amounts of time as the object goes around the Sun.

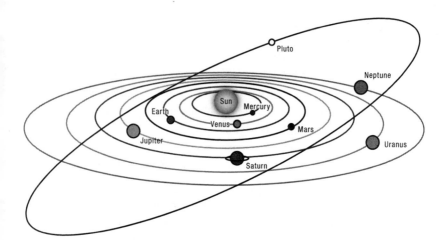

The worlds of the solar system shown on their orbital paths. Most of the planetary paths are close to being circular except for Pluto's orbit, which is very elongated.

Solar System Worlds and Orbits

PLANET	AVERAGE DISTANCE FROM SUN (KM)	ORBITAL PERIOD (EARTH YEAR/EARTH DAYS)
Mercury	57,900,000	0.24 Earth year (88 Earth days)
Venus	108 million	0.62 Earth year (226.3 Earth days)
Earth	149 million	1 Earth year (365.25 days)
Mars	227 million	1.88 Earth years (686.2 Earth days)
Jupiter	779 million	11.86 Earth years (4,380 Earth days)
Saturn	1.425 billion	29.5 Earth years (10,767.5 Earth days)
Uranus	2.85 billion	84 Earth years (30,660 Earth days)
Neptune	4.5 billion	165 Earth years (60,225 Earth days)
Pluto	5.06 billion	248 Earth years (90,520 Earth days)

What's a Planet?

Early Greek sky gazers used the word *planetes* (wanderer) to refer to starlike objects that wandered through the sky. Today, we apply the word *planet* to eight worlds of the solar system, excluding Pluto. In 2005, when planetary scientists found Eris, which is larger than Pluto, it forced them to think hard about what *planet* means. In the current definition (which will probably get revised again), a planet is defined by the International Astronomical Union (IAU) as a celestial body that has its primary orbit around the Sun, has sufficient mass for its own gravity to mold it into a round shape, and has cleared the neighborhood around its orbit by sweeping up all the planetesimals, which means that it's the only body of its size in its orbit. This complex definition excludes comets, asteroids, and smaller worlds that aren't rounded by their own gravity. The IAU also defined another class called *dwarf planets*. These are objects that meet the first two criteria for planets but have not yet cleared their orbits. Pluto, along with Ceres (discovered in 1801 and long known as a minor planet) and the more recently discovered worlds of Eris, Makemake, and Haumea, is now classified as a dwarf planet.

Forces That Sculpt Worlds

Several processes have an effect on the surfaces of worlds.

- *Volcanism* is when volcanoes spew mineral-rich lava. This happens on our own planet, Venus, and Jupiter's moon Io. It has occurred in the past on Mercury and Mars.
- *Cryovolcanism*, where icy material erupts from beneath the surface, occurs mostly on the frozen moons of the outer solar system.
- *Tectonism* warps the surface layers on a planet or moon, driven by heat from below. On Earth, tectonism stems from the motions of rock plates

that jostle around underneath our planet's crust. Tectonism may have occurred on Mars; it appears to have affected Venus; and a form of it occurs on some of the icy moons in the outer solar system.

- *Weathering* and *erosion* also change surfaces. On Earth, wind-driven sand can sculpt the landscape, and running water erodes the surface. This also occurs on Mars, where winds blow dust and sand across the surface. Extensive evidence shows that water once flowed across Mars's surface or existed in shallow seas and lakes.

THE SUN

Living with a Star

The Sun is a star and the biggest source of heat and light in our solar system. It's one of at least several hundred billion stars in the Milky Way Galaxy. Without it, life might not exist, and that makes it very important to us. To early people the Sun was something to worship. Ancient Greeks venerated Helios as the sun god, but they had insatiable scientific curiosity about it. They had lively debates over the true nature of this bright thing in the sky. In the 1600s, Italian astronomer Galileo Galilei (1564–1642) speculated about what the Sun could be. So did Johannes Kepler (1571–1630) a few decades later. In the 1800s, astronomers developed scientific instruments to measure the Sun's properties, which marked the beginning of solar physics as a scientific discipline.

Solar Physics

The study of the physics of the Sun is called *solar physics* and is a very active area of research. Solar physicists seek to explain how our star works and how it affects the rest of the solar system. They measure the Sun's temperatures, figure out its structure, and have assigned it a stellar "type" based on their measurements. Their work allows us to learn more about all stars by studying our own.

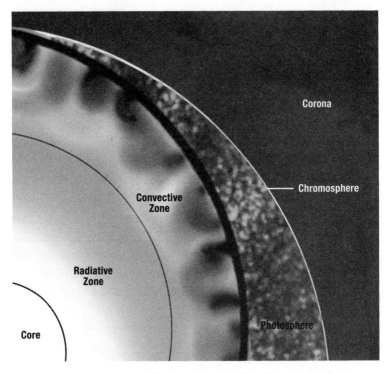

A cutaway view of the Sun, showing the core, the radiative zone, the convective zone, chromosphere, and corona.

Structure of the Sun

The Sun is essentially a big sphere of superheated gas. An imaginary voyage into its heart shows its structure.

- First, we have to traverse the outer solar atmosphere, called the corona. It's an incredibly thin layer of gas superheated to temperatures well over a million degrees.

- Once we're through the corona, we're in the chromosphere. It's a thin, reddish-hued layer of gases, and its temperature changes from 3,500° C (6,300°F) at the base to about 34,726 C (62,500°F) where it transitions up to the corona.

- Beneath the chromosphere is the photosphere, where temperatures range from 4,226°C (7,640°F) to 5,700°C (10,340°F). When you look at the Sun, the photosphere is what you actually see. The Sun is actually white, but it appears yellowish because its light travels through our atmosphere, which removes blue and red wavelengths from the incoming light.

- After we dive below the photosphere, we're in a layer called the convective zone. If you've ever boiled water or syrup, you might recall seeing little granular bubbles. The Sun's convective zone has these granular areas, too. They form as hot material from deep inside rises to the surface. These bubbles are actually currents moving through the Sun.

- The next layer down is the radiative zone. This is a very descriptive name because this region truly does radiate heat from the center of the Sun up to the convective zone.

- Finally, beneath the radiative zone is the solar core. This is the inner sanctum—a huge nuclear furnace. It's the place where nuclear fusion takes place. How does that happen? The temperature in the core is about 15 million degrees Celsius (27 million degrees Fahrenheit). The rest of the Sun pressing down on it provides a pressure 340 *billion* times Earth's atmospheric pressure at sea level. Those extreme conditions create a huge pressure cooker in which atoms of hydrogen slam together to form atoms of helium. The Sun fuses about 620 metric *tons* of hydrogen to helium each second, and that's what provides all of the heat and light.

Why the Sun Is Spotty

Sometimes the surface of the Sun is dotted with dark regions called sunspots. These are areas threaded with magnetic fields and look dark because they are cooler than the surrounding regions. Sunspots appear in eleven-year cycles. Halfway through each cycle, during a time called *solar maximum*, we see many sunspots. Then, they gradually taper off over five years until there are few sunspots. This is called *solar minimum*.

Sunspots are related to solar activity, particularly in bright outbursts called flares. Other solar explosions called *coronal mass ejections* hurl huge masses of energized gas (called *plasma*) out into space. These occur most often during the period of maximum solar activity and cause a phenomenon called *space weather*, which occurs throughout the solar system.

The Sun's influence extends throughout the solar system. Its heat and light travel out to the planets. Mercury is baked while the most distant worlds get a small fraction of the Sun's warmth. Our star also exerts another influence called the solar wind. This constantly blowing stream of charged particles extends out about 100 astronomical units (a hundred times the distance between Earth and the Sun) and creates a huge bubble that surrounds the solar system. The bubble's inner edge is called the heliopause. Beyond that lies interstellar space, where other stars go through the process of nuclear fusion just as our Sun does.

Did You Know?

If you could travel out to the heliopause, as the *Voyager* spacecraft has done, your view of the Sun would be very different from the friendly yellow star we see from our vantage point on Earth. It would be just one more star, set against a backdrop of millions of other stars. But, it's the star we live with, and that makes all the difference to us.

Exploring the Sun

Both professional and amateur astronomers study our star using observatories on Earth's surface and from space. Ground-based telescopes measure all aspects of the Sun's surface and atmosphere. Some radio telescopes and radar detectors track the Sun's influence on Earth's upper ionosphere. Their data is used in models that help predict space weather events. Solar physicists peer inside the Sun, using a set of special instruments called the Global Oscillations Network Group (GONG), which focus on sound waves moving through the Sun, a science called *helioseismology*. The Solar Dynamics Observatory is also equipped to do this work.

NASA has a fleet of space-based instruments that study the Sun and how it influences Earth:

- The Solar Terrestrial Relations Observatory (STEREO), two orbiting satellites that give a constant three-dimensional view of solar activity
- The Solar Dynamics Observatory (SDO), which gives real-time imagery of the Sun and its outbursts and does helioseismology
- The Solar Heliospheric Observatory (SOHO), which focuses on the Sun's upper layers and corona

SPACE WEATHER
The Sun-Earth Connection

Have you ever seen or heard about the northern or southern lights? These upper-atmosphere light shows are actually a very benign form of *space weather*, a term that describes changes in the near-Earth space environment mainly due to activity originating at the Sun. Such aurorae are upper atmosphere disturbances caused by solar activity. They occur about 80 to 90 kilometers (50–60 miles) above the polar regions of our planet. They are just one manifestation of a constant connection between Earth and the Sun. It's a connection that provides more than heat and light to our planet.

When Space Weather Attacks

The Sun constantly emits a stream of charged particles called the *solar wind*. As this wind rushes past Earth, it runs into our *magnetosphere*. That's the region of space around Earth that is bound by our magnetic field. Most of the wind slides right on by, but some of the charged particles get caught up in the magnetic field lines. They spiral in toward the polar regions since those areas are where the magnetic field lines originate. The charged particles energize molecules of gas in our upper atmosphere, called the *ionosphere*. This causes them to glow. That glow is called the aurora. If it appears over the north pole, it's called the *aurora borealis*; over the south pole it is called the *aurora australis*. Most of the time it glows white or green. However, if the solar storm is fairly energetic, more and different gases are energized and we can see reds and purples in spectacular auroral displays.

An Energized Sun

The Sun goes through periods during which it's more active. The most commonly known is the eleven-year sunspot cycle. It starts with a low number of sunspots and ramps up over five or so years to a point called *solar maximum*. That's when the Sun has the most sunspots and is usually cutting loose large outbursts of charged particles.

When the Sun sends a huge blast of charged particles in a giant outburst called a *coronal mass ejection*, or blasts out a giant *X-class flare* (the largest and most intense type of outburst), that material rushes out at high speeds. We see the light from the outburst 8.3 minutes after it happens. Masses of charged particles arrive anywhere from one to three days later, and they collide with Earth's magnetic field and stir up activity in the ionosphere.

Depending on how strong the flares and mass ejections are, we can experience anything from lovely auroral displays to much stronger and more serious consequences. A heavy solar outburst causes ionospheric disturbances called *geomagnetic storms*. These can affect or even shut down communications and global positioning satellites, endanger astronauts in near-Earth space, and in some severe cases, shut down power grids on our planet. How can that happen? A typical strong ionospheric storm stirred up by the collision of material from a strong solar outburst makes drastic changes in the near-Earth environment. Charged particles barrel right into our upper atmosphere. This creates strong electrical currents in the upper part of our atmosphere that affect ground-based electrical power grids.

It can disrupt our communications by interrupting the propagation of radio and other signals that people depend on. Satellites in near-Earth space are particularly vulnerable to the radiation from a space weather attack, as are the systems and people aboard the International Space Station. The threat of strong outbursts is driving serious research into predicting and preparing for damage to our technology. Currently our first line of defense is advance warning from a variety of orbiting satellites designed to study the Sun.

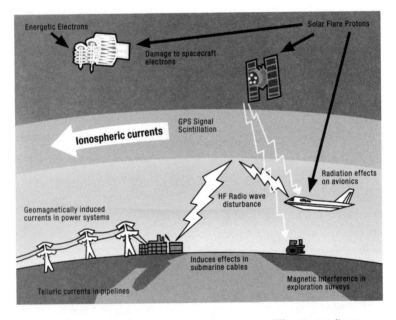

Space weather can affect Earth and our technology in many different ways. It can damage spacecraft electronics, cause power outages, and disrupt telecommunications.

Predicting Space Weather

Right now, no one can predict when the Sun will belch a mass of particles and radiation at us. Yet our civilization is highly dependent on a network of technologies that supply communication, transportation, navigation, and other modern services; these in turn are greatly affected by the Sun's outbursts. This is one important reason why atmospheric scientists and solar researchers take such a keen interest in studying the Sun's activity. Continued observations of outbursts provide valuable data about the process that ignites and delivers such things as X-class flares and coronal mass ejections. With that information, scientists can deliver timely warnings to satellite operators and power grid owners, as well as airlines, railroads, and other vehicle systems that depend on radar and GPS for navigation.

Space Weather and GPS

Global positioning satellite (GPS) data flows through our technological society. When you fly, take a train, cruise on a ship, or drive in a GPS-equipped car or truck, you're relying on satellite timing signals that travel through our atmosphere to receivers on the ground. That data helps calculate your position. Other systems also rely on GPS timing signals—these systems range from cellular telephones to financial transfer networks that move money around the globe. When a geomagnetic storm occurs, it increases the electron density (the number of electrons) in our upper atmosphere. Signals trying to propagate (move) through to the ground get delayed, which causes errors in their data. This means that systems depending on GPS have to shut down. If they're in planes, trains, or other modes of transportation, the operators must switch to other modes of navigation.

The Great Outage of 1989

On March 9, 1989, a coronal mass ejection lifted off the Sun headed directly toward Earth. Three days later, Earth experienced one of the most severe geomagnetic storms in recent history. It caused huge auroral displays, knocked out navigational controls on some satellites, and induced huge electrical currents that overloaded circuit breakers on parts of the Canadian power grid. The lights went out in Quebec for more than nine hours, leaving millions without power. This event, as well as others that occurred later that year, sent power grid owners on a quest to upgrade their systems so that such an outage wouldn't take them by surprise again.

MERCURY

A World of Extremes

Mercury is one of the most extreme places in the solar system. It's the closest planet to the Sun, and at 2,440 kilometers in diameter, it's the smallest planet in the solar system. It's actually smaller than Jupiter's moon Ganymede and Saturn's moon Titan. It is a dense, rocky world with a weak magnetic field generated from deep inside its gigantic core by the motions of molten material.

Mercury has a very thin, almost nonexistent atmosphere, and more than any of the other terrestrial planets, the landscape is cratered, cracked, and wrinkled. It's not hard to see why this world is so extreme, and that's what makes it so intriguing.

The Sun from Mercury

If you could stand on Mercury's surface, the Sun would look huge—three times bigger than from Earth. Because it's so close to our star, the planet's surface endures a temperature swing of nearly 610 degrees—the largest of any planet in the solar system. At its hottest, the surface bakes at 430°C (800°F). At night, temperatures drop to about -183°C (-300°F).

Mercury Facts

1. Closest point to Sun: 46 million kilometers (28.5 million miles)
2. Most distant point from Sun: 69.8 million kilometers (43.3 million miles)
3. Length of year: 88 Earth days
4. Length of day: 59 Earth days

5. Tilt of axis: 0.0 degrees

6. Gravity: 0.38 Earth's gravity

Mercury's History

Mercury, along with Venus, Earth, and Mars, formed from rocky materials that could withstand the high temperatures close to the newborn Sun. Each of these worlds has an iron-rich core, surrounded by a layer called the mantle. That mantle is made of the element magnesium, combined with minerals called iron silicates. Above the mantle on each planet is a rocky crust, created through volcanic eruptions and other geologic processes. Venus, Earth, and Mars all ended up with atmospheres above their surfaces, but early Mercury lost nearly all of its original blanket of gases due to constant buffeting from the solar wind.

Like the other planets of the inner solar system, Mercury was subject to impacts from incoming solar system debris during a period called the "Late Heavy Bombardment," which ended about 3.8 billion years ago. Most of the impact regions on Venus and Earth are gone, eroded away by atmospheric action, volcanism, and (on Earth) flowing water. Mercury, on the other hand, has retained many of the craters from that era of its history. Mars still shows some of its early bombardment, which tells us something about its later history. Generally, the more craters there are on a surface, the older a planet or moon is, and so Mercury's surface is quite old. In fact, in some areas we see impacts piled upon impacts, indicating those regions are quite ancient.

Some of Mercury's impact features, such as the 1,550-kilometer-wide Caloris Basin, were filled in with lava from subsequent eruptions. The object that created Caloris was probably a 100-kilometer-wide asteroid that slammed into the surface about 4 billion years ago.

Mercury Discoveries

Mercury turns out to be the densest of the terrestrial planets. It's made of about 60 percent iron—twice as much as Earth has—and 40 percent other minerals. Why it has so much iron is a mystery planetary scientists are working to solve.

To find out more about Mercury and the processes that shaped it, planetary scientists sent the *Mercury Surface, Space Environment, Geochemistry and Ranging (MESSENGER)* spacecraft on a multi-year mission of exploration. It arrived in orbit around the planet in 2011, and the data the spacecraft has returned are remarkable:

- There is strong evidence of water at the poles.
- Mercury's core is even larger than expected: It takes up 85 percent of the planet, is probably solid, and surrounded by a liquid layer of molten iron, a layer of iron sulfide above that, and silicate rocks that make up the crust.
- Images from the probe's dual imaging system and data from its laser altimeter instrument (which uses laser beams to measure surface features) show that Mercury had a very explosive volcanic early history. Massive lava flows erupted from vents and created long sinuous channels across the surface.

Volcanism isn't the only process that affected Mercury's surface. As the young Mercury cooled, it got smaller. Rocky crusts are quite brittle, and when the layers beneath them shrink, the surface rocks wrinkle and break. Mercury's landscape is split by large cliffs called scarps, which formed as the surface broke apart.

Mercury's geologic past isn't the only thing that *MESSENGER* is uncovering. The spacecraft's atmospheric spectrometer found a very tenuous off-and-on envelope of gas around the planet. This very thin atmosphere contains tiny amounts of hydrogen, helium, oxygen, sodium, calcium, potassium, and—surprisingly—water vapor. The source of most of the atmosphere comes from Mercury itself—its ancient volcanoes are still outgassing. Micrometeorites constantly slam down onto Mercury's rocky surface and kick up dust. In addition, the mission has measured particles of the solar wind in Mercury's atmosphere.

What about the water vapor? That comes from polar ice deposits. Scientists long thought that ice could exist where sunlight can't reach, and so they programmed *MESSENGER* to look for evidence. In 2012, instruments onboard the spacecraft detected hydrogen and other gases, suggesting that water ice and other volatiles (chemicals with low boiling points) exist at the poles. They also found traces of dark materials that could be organic compounds (carbon-based materials). It's likely those organics and the ice deposits were delivered by comets that smashed into Mercury's surface.

Missions to Mercury

There have only been two missions to Mercury since the dawn of the space age. *Mariner 10* was the first, visiting Mercury in 1973 and circling the planet three times. Its specialized instruments measured the planet and sent the first-ever close-up images of the surface. In 2011, the *MESSENGER* spacecraft inserted itself into orbit around the planet for a multi-year mission. There are more Mercury missions on the drawing boards, but no launches are planned until around the year 2015 at the earliest.

VENUS

Earth's Evil Twin

The planet Venus graces our skies so brightly that it has often been called the morning or evening star depending on when it appears. That brightness is due partly to the fact that Venus is the planet "next door" and because it is shrouded in clouds, which reflect sunlight. The planet's atmosphere is made mostly of carbon dioxide, a potent greenhouse gas that traps heat from the Sun. Hidden under Venus's clouds is a rugged desert.

The combination of a 462°C (863°F) surface temperature, an incredibly high atmospheric pressure some ninety-two times that of Earth's, and a history of volcanic eruptions has turned Venus into a hellish, alien landscape where no life could possibly exist. People often refer to Venus as Earth's twin because it's close to our planet's size and density. However, the nasty climate there makes it clear that Venus is really Earth's not-so-pleasant twin.

Venus Facts

1. Closest point to Sun: 107.4 million kilometers (66.7 million miles)
2. Most distant point from Sun: 108.9 million kilometers (67.6 million miles)
3. Length of year: 224.7 Earth days
4. Length of day: 117 Earth days
5. Tilt of axis: 177.3 degrees
6. Gravity: 0.9 Earth's gravity

The History of Venus

Venus is the second planet out from the Sun and is classified as a "terrestrial" planet. This means that it, like Mercury, Earth, and Mars, is made primarily of silicate rocks and metallic elements. It has a solid, rocky surface shaped by volcanic activity that may still be occurring, tectonism (earthquakes), erosion, and weathering. These processes also work to modify the surfaces of other planets and some of the moons of the solar system.

Early in its history Venus may have had water on its surface, but for some reason the planet lost it very quickly and developed the thick carbon dioxide atmosphere we see today. Volcanism continued to resurface early Venus, flooding the impact craters created during a period called the "Late Heavy Bombardment" (which ended about 3.8 billion years ago). This was a period when the worlds of the inner solar system were particularly targeted by debris left over from the formation of the solar system.

Something also stopped Venus from developing plate tectonics— the large-scale motions of the outer crust of a world—and planetary scientists are still working to figure out more of Venus's evolutionary history. There's not a lot known about the interior of Venus, but the best models suggest that it probably has a partially molten core, surrounded by a mantle layer and a crust. The mantle absorbs heat from the core, and when this middle layer gets too hot, it weakens. This causes the top layer—the crust—to melt in on itself, spurring volcanic events that keep paving the surface over and over again.

The planet also has no internally generated magnetic field. However, there is a weak field that seems to be created as the solar wind interacts with Venus's upper ionosphere. The European Space Agency's *Venus Express* mission has observed interesting *flux ropes* high in the atmosphere above the poles. Other planets with stronger magnetic fields have these ropes, but they're rare for Venus. Flux ropes are magnetic structures that arise in a magnetic field when a high-speed solar wind stream is flowing past a planet.

Craters on Venus

Venus has a surprising number of craters despite the volcanic activity that works to erase them over time. There are nearly a thousand craters ranging in size from about 4 to 280 kilometers across. The size of the craters imply that only objects larger than 50 meters across make it through the atmosphere to smash into the surface. Smaller impactors are simply vaporized in the heavy atmosphere.

The Runaway Greenhouse

Early in its history, the wet, temperate Venus that some scientists think existed began to change. The most likely explanation is that as the newborn Sun brightened up, it heated the early Venusian atmosphere. This created water vapor. Eventually the oceans boiled away, and all the water vapor escaped out to space. The carbon dioxide atmosphere remained, along with the sulfuric acid clouds that shroud the surface from our view.

Based on radar studies from Earth and multiple spacecraft missions to Venus, today we know that the planet's upper atmosphere circulates around the planet in only four Earth days while the planet itself turns much more slowly. There are also intriguing double-eyed hurricane-like storms called polar vortices, which whip along at a speed sixty times faster than the planet below it.

Venus is sometimes cited as an extreme example of what could happen to Earth as our planet's atmospheric carbon dioxide buildup continues. Nothing so drastic will happen to Earth, although the rise in greenhouse gases here is troubling. Understanding the Venusian greenhouse helps climate scientists see what *can* happen to an atmosphere that is made largely of carbon dioxide and smothered in heavy clouds.

Exploring Venus

Humans have been observing Venus for centuries. First we used telescopes to explore this cloudy world. Then, the space age offered us a way to go to Venus and study it up close. There have been thirty-eight missions sent from the United States, the former Soviet Union and Russia, the European Space Agency, and Japan to study Venus. Not every one has been successful—some were plagued with instrument failures and others were lost.

However, there have been a solid string of successful Venus explorations, including:

- The *Venera 4* and *5*, the first to enter the atmosphere and send information
- *Venera 9*, which sent back the first images from Venus
- The *Magellan Mission*, a thirteen-year-long Venus radar mapping project
- The *Pioneer Venus* orbiters
- The current *Venus Express*

All of these missions provided long-term observations of the atmosphere.

Observing Venus

Venus is one of the most often-observed objects in the sky. Each month, there are people who see it and think that it's a bright star or a UFO. It appears either in the morning skies before sunrise or in the evening skies right after sunset. You can also sometimes spot Venus during the day.

If you have a telescope, watch Venus over a period of weeks as it makes its way around the Sun. Its apparent shape seems to go through phases, much like the Moon does. It ranges from a small, full appearance to a quarter phase when it is farthest away from the Sun in its orbit.

EARTH

The Home Planet

In the late 1960s, the *Apollo* missions to the Moon gave us our first look at Earth as a planet seen from outer space. In 1990, the *Voyager 1* spacecraft beamed back a "family portrait" of the entire solar system from its vantage point some 6 billion kilometers (3.7 billion miles) away. Earth appears as a tiny, pale blue dot only a few pixels across. That shot was requested by astronomer Carl Sagan, and in his book *Pale Blue Dot*, he pointed out that this tiny dot is all we know. He wrote, "On it everyone you love, everyone you know, everyone you ever heard of, every human being who ever was, lived out their lives."

Earth, for now, is also the only world on which we know that life exists. It lies in a safe warm zone around the Sun, and is home to nearly 9 million known species of life today (with many more yet to be discovered). It has hosted many times that number in the past.

Earth Facts

1. Closest point to Sun: 147.0 million kilometers (91.4 million miles)
2. Most distant point from Sun: 152.0 million kilometers (94.5 million miles)
3. Length of year: 365.25 days
4. Length of day: 23 hours, 56 minutes
5. Tilt of axis: 23.5 degrees

Earth is a unique, watery world with an evolutionary history that we're still coming to understand. Its tilt, coupled with its orbit around the Sun, gives us distinct seasons throughout the year. Scientists study our weather, climate, surface features, and many other aspects that help explain our planet's formation and place in the solar system. What we learn about our planet teaches us about geology, marine biology, oceanography, atmospheric science, paleontology, and many other "ologies" that describe our home world and help us learn more about other planets.

Earth's History

Our planet began life as a small, rocky object in the center of the protosolar nebula (the solar system's birth cloud) about 4.5 billion years ago. Like the other inner planets, it coalesced through collisions of smaller rocky bodies (a process called accretion) to become more or less the size it is today. The infant Earth was a molten world, continually bombarded by material in the cloud of gas and dust surrounding the newborn Sun. As it cooled, it experienced constant eruptions, and it had an atmosphere of noxious gases created by outgassing from volcanoes. It didn't form with a natural satellite, but one came into being when Earth and an object about the size of Mars collided about somewhere between 30 and 50 million years after the solar system began to form. The debris from that event coalesced to form the Moon we know today.

As the young Earth cooled, it formed a solid crust that became the continental plates. They ride on top of a mantle layer. Beneath the mantle is the core, which is divided into two parts: the inner and outer core. This formation of layers is called *differentiation*, and it didn't just happen on Earth. Other planets have differentiated layers, as do dwarf planets, the Moon, and some asteroids.

After the period of bombardment and frequent collisions ended about 3.8 billion years ago and Earth's crust began to cool, the oceans started to form. In fairly short order, the first forms of life appeared. They began to fill the atmosphere with oxygen, and over time our blanket of gases was able to support more diverse life forms, including us.

Earth will exist, along with the other inner planets, for another 5 billion years. That's when the Sun will begin to swell up to become a red giant, heating up the solar system even more. At that point, Earth's oceans will boil away, and the planet will become a lifeless cinder.

Earth's Atmosphere and Oceans

As with other planets that have atmospheres, Earth's is a mixture of gases.

Composition of Earth's Atmosphere

Nitrogen	78.084%
Oxygen	20.9%
Argon	0.9340%
Carbon dioxide	.0394%
Neon	0.0018%
Helium	0.0005%
Methane	0.00018%
Krypton	0.00011%
Hydrogen	0.00005%
Nitrous oxide	0.000032%
Other elements: 0.079 plus water vapor.	

The atmosphere is a protective blanket, absorbing most solar ultraviolet light. It also keeps temperatures warm through what's called the greenhouse effect. Heat from the Sun is absorbed by gases such as carbon dioxide, which radiate that heat to the Earth's surface.

The greenhouse gases are what make life livable on our planet. However, we are now affecting our atmosphere through the release of high levels of greenhouse gases. This is affecting global temperatures, speeding up the melting of Arctic ice and warming the oceans.

Nearly three-quarters of our planet is covered with water in the form of the global ocean, lakes, and rivers. This is called the *hydrosphere*. The oceans influence long-term climate patterns and short-term weather changes, and they are a principal part of the carbon cycle—the method that our planet uses to exchange carbon between the atmosphere, the oceans, and the surface.

The oceans are the last frontiers for exploration on our planet, and oceanographers estimate that only about 5 percent of the ocean bottom has been explored. Undersea volcanoes, mountain ranges, and basins are hidden from our view, yet these are an important part of Earth's geology.

Where Did the Oceans Come From?

When Earth formed, it had no seas. Where did they come from? One theory says that the oceans were delivered as part of a vast bombardment of icy bodies called cometary nuclei. They were especially plentiful during planetary formation in the inner solar system and could have easily collided with the newborn Earth. However, some scientists argue that there had to be a homegrown source of water here. The same protoplanetary debris that accreted to form Earth also contained water and water ice. So, it's possible that the oceans came from water that was already part of the Earth's rocky components as it formed.

Life's Origins on Earth

How did life get its start here? Where did it begin? It's hard to answer these questions precisely, but it's clear the earliest living beings arose from chemical origins. Some suspect it began as mats of organic molecules in shallow ponds. Others suggest that life, which needs water, warmth, and organic (carbon-containing) material to survive, began in the deep oceans around volcanic vents. Still others think that complex organic molecules in our atmosphere were energized by lightning strikes, which sent them down the road toward life. While scientists are still working to find the answer to the question of how life got started here, the models agree that life arose out of chemical elements that combined in certain ways. All it took was a good location, water, some energy, and time.

THE MOON

Earth's Natural Satellite

The Moon has fascinated humans since the dawn of time. It is our nearest satellite, and has the distinction of being the only other world where humans have walked. Beginning in 1969 the *Apollo 11–17* missions took astronauts to the Moon. They spent time doing scientific studies of the Moon's surface. They brought back a treasury of rocks and dust that helped researchers understand the origin and evolution of our closest natural satellite.

- Number of past lunar missions: 10
- Number of astronauts who flew to the Moon: 24
- Number of astronauts who walked on the Moon's surface: 12

The Moon's Changing Face

The Moon's appearance in our skies changes over time. Those changes are called lunar phases, and they occur over a period of twenty-nine and a half days, beginning with New Moon and going through quarter moon, slim crescent, and Full Moon before returning to New Moon again. The Moon orbits Earth once every twenty-seven days, and it always shows the same face to us. This is because it is locked into synchronous rotation, which means that it takes as long to rotate once on its axis as it does to orbit Earth.

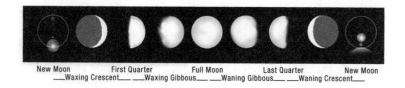

New Moon   First Quarter   Full Moon   Last Quarter   New Moon
___Waxing Crescent___ ___Waxing Gibbous___ ___Waning Gibbous___ ___Waning Crescent___

Depending on where the Moon, Sun, and Earth are located in relation to each other, we see a different phase of the Moon throughout the month.

The Surface of the Moon

When you look at the Moon, it shows dark and light areas. Those dark areas are often referred to as *maria*—the plural form of the Latin word *mare*, which means "sea." Early Moon watchers thought those regions were watery oceans, but a close look with a telescope or a pair of binoculars shows no water on the surface, just rocky plains. The low-altitude maria formed as volcanic vents called *lunar domes* emptied out their molten basaltic lava and flooded the surface. The light areas are called the *lunar highlands*. They are mostly hilly regions that lie at higher altitudes than the maria. The whole surface is peppered with impact craters, made as solar system debris crashed into the Moon.

Water on the Moon

Although there are no oceans of water on the Moon, there *is* water there. Several missions have found the chemical fingerprints of water, indicating that water either exists in rocks (chemically bound to minerals) or there may be ice deposits near the polar regions. The ice deposits could have been delivered during impacts of comets and water-bearing asteroids and meteoroids. The discovery of water on the Moon is important: If enough can be found, it would be useful as humans plan long-term colonies and research bases.

If You Could Go to the Moon

The Moon has a very weak gravitational field only one-sixth as strong as Earth's. This means that if you were a 200-pound person exploring the Moon, you would weigh only about thirty-two pounds there. As the astronauts did during the *Apollo* missions, you would have to wear a space suit to supply water and oxygen since the Moon has no atmosphere and no liquid water on its surface. This also means that your suit would need to shield you from the Sun and radiation hazards. People living on the Moon would probably live *in* the Moon, in underground cities sheltered from the harsh surface environment.

Inside the Moon

The interior of the Moon is made up of several layers. It's a differentiated world because it consists of layers, each having a slightly different mineral makeup. The surface is covered with a thick layer of dusty material called *regolith*. Below that is the crust, and it's made mostly of a mineral called *plagioclase*, which is also found on Earth. The crust ranges in thickness from 60 to 150 kilometers (37 to 93 miles). Beneath it is the mantle, which is made of such iron-rich minerals as olivine, which is also plentiful on Earth. The central part of the Moon—the core—may be partially molten and is probably made of iron-rich materials. The central part of the core is almost certainly pure iron. The Moon experiences "moonquakes" fairly often. They are thought to be caused by gravitational interactions with Earth, by impacts, or when the surface freezes and thaws.

Lunar Fact

You often hear the adjective *lunar* used when referring to the Moon's surface or structure, or missions to the Moon. That term comes from the Latin word *Luna*, one name for the Roman goddess of the Moon.

How the Moon Formed

The Moon very likely formed about 4.5 billion years ago in a titanic collision between the newborn Earth and a Mars-sized object referred to as Theia. The impact threw huge amounts of material out into space, and eventually much of it was pulled together by gravity and collisions to form the Moon. The heat of the collision and subsequent accretion melted surface rocks, creating a magma ocean on the newborn Moon. Interestingly, based on chemical analysis of Moon rocks brought back by the *Apollo* astronauts, the rocks on the Moon have characteristics identical to Earth rocks. This means that most (if not all) of the materials that formed the Moon originated as part of Earth.

Current and Near-Future Moon Missions

NAME	MISSION GOALS	COUNTRY	STATUS
Lunar Reconnaissance Orbiter	Surface mapping	USA	Working
Chang'e 2	Lunar studies, now studying asteroid Toutatis	China	Working
LADEE	Study lunar atmosphere and surface	USA	Launch in 2013
Chang'e 3	Soft lander, perform UV astronomy	China	Launch in 2013

Current and Near-Future Moon Missions

NAME	MISSION GOALS	COUNTRY	STATUS
GLXP	Moon Express lander, rover	Private competition	Launch ~2013
Luna-Glob	Robotic lunar base/orbiter combo	Russia	Launch ~2014
Chandrayaan-2	Robotic rover, orbiter combo	India	Launch ~2016

Living and Working on the Moon

People are interested in getting back to the Moon for a number of reasons. One of the biggest is that it is a scientifically interesting place. It can help us understand more about its formation and Earth's history. It may be a treasure trove of minerals and other materials that could be used to construct future bases and plan solar system exploration missions. The Moon also presents a perfect place to do astronomy. During its thirteen-day "night" period, astronomers could aim telescopes to study dim and distant objects without having to worry about light pollution. Radio astronomers are also excited about having installations that are not disturbed by radio interference from Earth.

No matter what work people do on the Moon in the future, they will have to contend with a much harsher and different environment than the one we currently live and work in here on Earth.

MARS

Unveiling Red Planet Mysteries

Mars—the Red Planet, fourth rock from the Sun, representing the god of war in at least a dozen cultures—has fascinated humans throughout history. At first glance, it looks Earth-like. There are mountains, volcanoes, canyons, vast plains, polar ice caps, and a pale, sometimes cloudy sky that changes from pink or red at sunset and sunrise to a yellowish brown during the day. However, there are some things missing from Mars: surface water, a thick atmosphere, and life. Also, it has impact craters, and while Earth does have a few, Mars has many more. The red surface color, due largely to the presence of iron oxides (rust) in the soil, is another clue that Mars, no matter how familiar its landscapes may look, is *not* Earth. It's a very different world, one with a past planetary scientists are working to understand. Its two moons, Phobos and Deimos, pose geological mysteries of their own. They were likely asteroids, captured by Mars's orbit sometime in the distant past.

Mars Facts

1. Closest point to Sun: 206.6 million kilometers (128.4 million miles)
2. Most distant point from Sun: 248.2 million kilometers (154.8 million miles)
3. Length of year: 1.88 Earth years
4. Length of day: 24 hours, 37 minutes
5. Tilt of axis: 25 degrees
6. Gravity: 0.37 Earth's gravity

Mars: From the Past to the Present

Mars formed 4.5 billion years ago, about the same time as Earth. The newborn planet cooled quickly and formed a thick crust, but no tectonic plates. The core also cooled, which stopped the dynamo action and left Mars with no way to generate a strong magnetic field. Early Mars had a warm, wet, and thick carbon dioxide atmosphere, but that changed about the time the magnetic field disappeared. Ancient floodplains, riverbeds, and shorelines of ancient lakes and shallow oceans are now seen everywhere. Along with layered sedimentary rock, they suggest the existence of liquid water in the past.

For a long time the lack of water on Mars has been a mystery. Where did it and the atmosphere go? Billions of years ago, the combination of Mars's low gravity and lack of magnetic field simply allowed much of the atmosphere to escape into space. That caused the temperatures and atmospheric pressure to drop below the point where liquid water could exist. What's left of Mars's water is locked up in the polar caps and ice deposits below the dusty surface. On Earth when subsurface ice melts, it shapes some spectacular landscape changes. The same happens on Mars. In the north polar region, spacecraft images show what's called patterned ground where ice has frozen, softened, and then remelted. In other places where subsurface ice melted, the overlying layers of rock settled in, forming what is called chaotic terrain. The water released during the collapse carved out large flow channels.

Mars also experienced tremendous volcanic activity. The largest of its volcanoes—Olympus Mons—rides high atop a thick piece of the crust called the Tharsis Bulge. As the volcano and the bulge built up over time, tectonic stresses cracked the surface. The famous Valles Marineris is one such crack. It stretches across a third of the planet.

This system of canyons has also been carved by wind erosion, and there is some evidence that parts of it were eroded by flowing water.

Why Study Mars?

The most important reason why we keep sending missions to Mars is to search for evidence of life. If Mars was wet and warm in the past, then it had the necessary ingredients for life: water, warmth, and organic material. So, now the questions are: Did life form on Mars in the past? If so, what happened to it? Does it exist there now? If so, where? What is it? The missions to Mars are a way to search for much-needed answers.

Today, Mars is a dry and barren desert planet. Its thin atmosphere is made mostly of carbon dioxide with small amounts of nitrogen and traces of argon, oxygen, and water vapor. The atmospheric pressure on Mars is very low, about 6 percent of Earth's surface pressure. The planet is much less massive than Earth, and that gives it a gravitational pull about 30 percent less than ours.

The combination of the tilt of its axis and its lengthy year (almost twice as long as Earth's) gives Mars a seasonal climate. Each season is roughly twice as long as Earth's. A balmy day on Mars in mid-summer at the equator might get up to 35°C (95°F) according to measurements taken by the NASA *Mars Opportunity* rover. Most of the time Mars is quite a bit colder, perhaps closer to 0° C. On a cold wintry night, the temperature can plummet to -150°C (-238°F). Images of Mars taken in the colder months show frost covering the ground, much as it does when the weather turns very cold here on Earth. Future Mars explorers will have to wear space suits that help keep them warm and pressurized and supply oxygen to breathe.

Exploring Mars

After Earth and Venus, Mars is one of the most-explored planets in the solar system. Telescopes around the world continually monitor the planet, and the *Hubble Space Telescope* keeps a watch from orbit. Space agencies from the United States, the former Soviet Union, Japan, and the European Space Agency have sent dozens of spacecraft to Mars since the early 1960s. At last count, seventy-five missions have gone, and NASA is funding a new Mars rover project to go to Mars in 2020. The Indian Space Research Organization is also working on the *Mangalyaan* Mars mission.

Recent successful missions include:

- The *Mars Odyssey* Orbiter (launched April 7, 2001)
- The Mars Exploration Rover *Spirit* (landed January 4, 2004)
- The Mars Exploration Rover *Opportunity* (landed January 25, 2004)
- The *Mars Reconnaissance* Orbiter (launched August 12, 2005)
- The Mars Science Laboratory Rover *Curiosity* (landed August 6, 2012)

These advance scouts have sent back dazzling images, studied the dust and rocks searching for water and life residues, and made atmospheric measurements. In the future, the *Mars Atmosphere and Volatile Evolution* (MAVEN) mission will help scientists figure how and when Mars lost its atmosphere and surface water. In 2016 and 2018, the European Space Agency and the Russian Roscosmos agency will launch the *ExoMars* mission to search for traces of life called *biosignatures*.

Mars in Popular Culture

Mars has symbolized many things in human history. It started out as the god of war and bloodshed, mainly due to its blood-red color. That warrior-like characteristic has persisted to modern times. The *War of the Worlds*, a science-fiction novel by H. G. Wells (1866–1946), portrayed Martians as bloodthirsty invaders. Science fiction writer Edgar Rice Burroughs (1875–1950) imagined all manner of aliens battling it out on Mars. However, writer Robert A. Heinlein (1907–1988) saw Martians as peaceful "Old Ones." The popular 1996 movie *Mars Attacks* spoofed the Hollywood fascination with aliens, using invading Martians as the bad guys. Today, we have Mars landers that "send" Twitter messages and whole websites devoted to Mars missions, where enthusiasts can learn more about the ongoing exploration of the Red Planet.

JUPITER

The King of the Planets

Jupiter is a world of superlatives. It's the largest planet in the solar system. More than eleven Earths would fit across its diameter. It's also the most massive. More than 1,300 Earths could fit inside Jupiter, with room to spare. This gas giant is also a weird place on the inside. The interior consists of a series of metallic and liquid hydrogen layers covering an Earth-sized, partially rocky core. Jupiter takes twelve years to go once around the Sun, and it lies more than five times farther from the Sun than Earth does.

Jupiter Facts

1. Closest point to Sun: 740 million kilometers (460.2 million miles)
2. Most distant point from Sun: 816 million kilometers (507 million miles)
3. Length of year: 11.8 Earth years
4. Length of day: 10 hours
5. Tilt of axis: 3.13 degrees
6. Gravity: 2.64 Earth's gravity

Between Jupiter's fast rotation (ten hours) and the action of metallic liquid hydrogen deep inside, this planet has the strongest magnetic field in the solar system. Because of its mass, Jupiter's gravitational pull at its cloud tops is more than two and a half times Earth's gravity at sea level.

Jupiter has the largest planetary atmosphere of any object in the solar system. It's composed of hydrogen and helium, and its uppermost regions contain three layers of clouds. One is made of mostly ammonia ice crystals; another is a mixture of ammonia and sulfur; and the third one has clouds of water vapor. The top layers are divided into belts and zones, and winds whip through these regions at speeds over 600 kilometers (372 miles) per hour. That action spawns huge storms and vortices, including the largest in the solar system—the Great Red Spot.

The Galilean Moons

Jupiter is orbited by at least sixty-three known moons. The four largest are Io, Europa, Ganymede, and Callisto (in order of their distance from the planet). They were discovered in 1610 by astronomer Galileo Galilei and are called the *Galilean Moons* in his honor.

Io is the closest one to Jupiter and is a volcanic world, spewing so much lava onto its surface that it may have literally turned itself inside out over millions of years. Sulfur plumes shoot out from those volcanoes, and eventually their material ends up in orbit around Jupiter. Why is Io so active? It's caught in a gravitational tug-of-war between Jupiter and the combined gravitational pull of the other three Galilean satellites. The tugging distorts Io's shape, and this causes friction heating inside, which powers the volcanic activity. This is called tidal heating and affects the interiors of other moons in the solar system, too.

Europa is a water-rich world with a thin, icy crust and a tenuous atmosphere consisting mostly of oxygen. There's very likely a deep ocean of water heated by the decay of radioactive elements in this moon's core. It could support some form of life, since conditions are so favorable there. However, this little moon experiences deadly radiation due to its proximity to Jupiter, and its position inside the planet's magnetic field could affect any life that managed to evolve there. Scientists will find out more if and when a mission to Europa is ever sent to explore what lies beneath its surface.

The other two Galilean moons—Callisto and Ganymede—are fascinating in their own rights. Ganymede is the largest satellite in the solar system and larger than the planet Mercury. Its surface is dark and crossed with grooves and ridges. Here and there are white patches where pieces of solar system debris ploughed into the crust and spattered ice out from the craters they created. Interestingly, Ganymede has a weak magnetic field generated by activity deep inside, and a thin atmosphere that contains oxygen and possibly some ozone and hydrogen.

Callisto is a dark-looking moon with a very old surface spattered with ancient impact craters. It doesn't seem to have any internal activity, as the other Galileans do, although it could have a watery ocean beneath its icy crust.

Jupiter's Ring System

In 1978, cameras aboard the *Voyager 1* spacecraft imaged a thin set of rings circling Jupiter. It's no wonder no one saw them before—they're very faint and dusty, not bright and icy like Saturn's. The *Galileo* spacecraft mapped them in great detail, and the *Hubble Space Telescope* has observed them. The rings are divided into several sections: a halo ring very close to the planet, the main ring, and a pair of gossamer rings that consist mostly of dust ejected from the moons Adrastea, Amalthea, and Thebe.

Exploring Jupiter

Galileo Galilei touched off Jupiter exploration with his first glimpse of it through a telescope in 1610. Nowadays, ground-based telescopes, including good amateur instruments, as well as the *Hubble Space Telescope* can make out the planet's four large moons and its cloud belts and zones. In 1994, observers watched in real time as twenty-one pieces of Comet Shoemaker-Levy 9 plowed into Jupiter and left behind dark smudgy scars in the upper cloud decks. More collisions were detected in 2009, 2010, and 2012. This led astronomers to nickname Jupiter "the solar system's vacuum cleaner" because its strong gravity attracts and deflects comets and other solar system debris.

To really get to know Jupiter, planetary scientists had to get up-close and personal. Their first mission was the *Pioneer 9* spacecraft, which flew by the planet in 1973, returning high-resolution images of the cloud tops and gathering data about the magnetic fields and upper atmosphere. Since that time, the planet has been visited by *Pioneer 10, Voyagers 1* and *2, Galileo, Ulysses, Cassini,* and *New Horizons.* The *Galileo* mission was the first (and so far only) mission to go into orbit around Jupiter. It arrived in December 1995 and spent more than seven years gathering data about the planet, its magnetosphere, and its moons. The *Juno* mission will arrive at Jupiter in 2016 and begin a five-year mission to study the planet and its environment, using specialized instruments to study thermal radiation (heat) coming out of Jupiter's interior, measure its gravitational and magnetic fields, and study its auroral displays (similar to northern and southern lights on Earth). Other missions to Jupiter are on the drawing board, including ESA's *Jupiter Icy Moons Explorer (JUICE)*, which should be launched in 2022. Its planners hope to spend three years studying the Jovian system and its moons. The continued exploration of this giant planet will help explain more about its origin and evolution in the solar system.

The Great Red Spot

The winds that blow through Jupiter's upper cloud tops stir up storms that look like cyclones and typhoons here on Earth. The Great Red Spot anticyclone has been whirling through the upper atmosphere for at least 350 years and was first observed in 1665 by astronomer Giovanni Dominico Cassini (1625–1712). This storm is so big that three Earths could fit comfortably inside it. The Great Red Spot may get its colorful appearance from a mixture of red phosphorus, possibly some sulfur, and a scattering of other organic compounds. It has been known to turn a salmon pink and also to lose most of its color altogether for short periods of time.

SATURN

The Original Ring World

Saturn is one of the most popular planets to view through a telescope. Even a small one shows off the planet's beautiful ring system, as astronomer Galileo Galilei found out when he first spotted it through his low-power instrument in 1610. He sketched what he saw and couldn't imagine what made Saturn look like it had ears or maybe a pair of moons on either side. It took nearly 250 years to determine the nature of these appendages. In 1655 Dutch astronomer Christiaan Huygens (1629–1695) looked through a more powerful telescope and decided that Saturn had a disk of material around it. Giovanni Domenico Cassini (1625–1712) in 1675 determined the ring to be composed of multiple rings with gaps between them. The true nature of the rings as a system of small particles in orbit around Saturn was proposed by James Clerk Maxwell (1831–1879) in 1859. Eventually, when *Voyagers 1* and *2* and the *Cassini* mission explored this world, their images showed myriad ringlets, with gaps containing dust and small embedded moonlets that cause complex patterns in the rings such as waves, wakes, and other transient features.

The classical ring system extends out to about 121,000 kilometers (75,189 miles) away from Saturn. There are several notable gaps such as the Cassini Division in the rings. Two of the gaps are caused by moons sweeping through and clearing a path, while others are the result of gravitational interactions between Saturn and its other moons. Beyond the classical rings lie narrower ringlets, such as the G and E rings, which extend out to the orbits of several other smaller Saturnian moons.

The world beneath those rings is a gas giant planet that is so massive and voluminous that 763 Earths could fit inside. Like most other planets, Saturn is a layered world. It has a rocky core smothered in layers of liquid metallic hydrogen (where electrical currents stir up activity and generate its rather weak magnetic field) and liquid helium. Its atmosphere is made of hydrogen and a small amount of helium, plus traces of methane, ammonia, ethane, and other organic materials. The top of Saturn's atmosphere is covered with a layer of ammonia crystal clouds, which give it a pale, muted color. Saturn itself rotates on its axis once every ten and a half hours, a speed so fast that it flattens the planet and makes it the most oblate of all the planets.

Saturn Facts

1. Closest point to Sun: 1.3 billion kilometers (838.7 million miles)
2. Most distant point from Sun: 1.5 billion kilometers (934.2 million miles)
3. Length of year: 29.4 Earth years
4. Length of day: 10 hours, 39 minutes
5. Tilt of axis: 26.73 degrees
6. Gravity: 0.92 Earth's gravity

Most of Saturn and its upper atmosphere rotate at different rates, and this causes winds to blow at speeds up to about 1,800 kilometers (1,100 miles) per hour. These winds produce some bands in the clouds, but they're hard to see through the hazy clouds above them. Occasionally storms pop up, whirl their way through the clouds, and then dissipate. One of the most unusual features in the Saturnian atmosphere is the North Pole hexagonal cloud pattern. There's a vortex at the heart of the pattern that looks exactly like the eyewall of a hurricane here on Earth. It's still not clear why this vortex exists, and astronomers are using *Hubble Space Telescope* and the *Cassini* orbiter to understand more about it.

The Birth of Saturn

Like the other planets in the solar system, Saturn formed some 4.5 billion years ago in a cloud surrounding the newborn Sun. The areas of the cloud closer to the Sun contained primarily rocky particles and other heavier elements—which became part of the planets Mercury, Venus, Earth, and Mars. The outer solar system, where temperatures were colder, was more hospitable to planetesimals (the seeds of planets) that were rich in volatiles (gassy materials) and ices. Those small worldlets coalesced together to create each of the gas giants. The gravity of each of the new worlds was so heavy that it swept up much of the remaining material, which became the hydrogen layers and atmospheres of the gas giants. Saturn itself cooled down rather quickly after it formed, and that helped build its helium and hydrogen interior layers. The inner regions still give off more heat than the planet receives from the sunlight that shines onto its cloud tops. That heating contributes to the stormy weather that occurs in the cloud tops.

Saturn's Fabulous Moons

Not only does Saturn have a spectacular ring system, but it is also accompanied by a retinue of sixty-two known moons. The major ones are:

- Titan
- Mimas
- Enceladus
- Tethys

- Dione
- Rhea
- Iapetus

These are all worlds made primarily of ice with some rock mixed in under their surfaces. They and many other Saturnian moons probably formed as part of a smaller nebula that surrounded their parent planet during the formation of the solar system.

Mysterious Titan

When the *Huygens* lander settled onto the surface of Titan in 2005, it revealed a surface of frozen methane ice and methane lakes shimmering in a nitrogen atmosphere fogged with clouds made of ethane and methane. Titan undergoes seasonal weather changes, and periodic rains and winds create dunes, flowing rivers, and small seas and lakes. Beneath the surface is a world made mostly of water ice and rock. This may sound like a very un-Earth-like environment, but it turns out that early Earth may have been very similar to what Titan is today. There has been no life spotted on Titan, but chemical analysis of the atmosphere may yet reveal whether it ever did form there. If it did, such life would be chemically very different from life as we know it.

Exploration of Saturn

In addition to Earth-based exploration of Saturn, four spacecraft have visited the planet. *Pioneer 11* and *Voyagers 1* and *2* flew by the system in 1979, 1980, and 1981, respectively. The *Cassini Solstice Mission* went into orbit in the Saturnian system in 2004. In 2005 the *Huygens* probe landed on cloud-shrouded Titan and provided a first look at its frozen surface. There are currently no plans for further missions to Saturn, but the data and images returned by the *Voyager* and *Cassini* missions are providing work for years of future study of the ringed planet and its moons.

Enceladus: A Moon of Geysers

Enceladus is a fascinating place. It's a frozen world with an icy crust and liquid water hidden deep below. That water gets forced to the surface, where it vents out through huge geysers that shower ice particles into orbit as part of Saturn's rings. Some of those ice crystals fall back to Enceladus as a fine snow. This venting process is called *cryovolcanism*, and it takes place on several other ice moons in the outer solar system. The *Cassini* spacecraft first found the Enceladus geysers in 2005, and the spacecraft went through the plume of gas and found water vapor and small amounts of nitrogen, methane, and carbon dioxide. Because it has a deep ocean of water and heating in its central core, Enceladus is considered a good place to look for evidence of life.

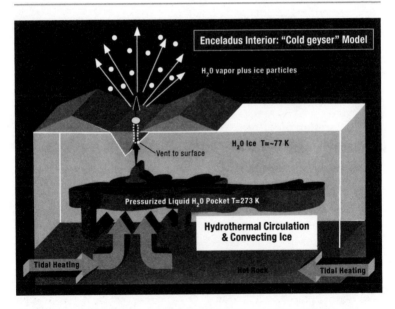

This cross-section shows one possible explanation for the geysers shooting out material from beneath the surface of Enceladus. Pressurized liquid water may be forced out to the surface by heat from the core of this tiny moon.

URANUS

The Tipped-Over Giant

Greenish-blue, haze-covered Uranus is well known in the planetary science community as the planet that orbits the Sun on its side. None of the other planets do this. During its orbit, one pole or the other on Uranus points toward the Sun during the solstices. At the equinoxes, the planet's equator points toward the Sun. This combination of axial tilt and the lengthy year gives Uranus some interesting day-night scenarios. For example, each pole gets forty-two Earth years of sunlight. During the equinoxes, the equatorial regions have very short days and nights because the Sun never rises very high over the horizon. Uranus is made of a higher percentage of water, methane, and ammonia ices than Jupiter and Saturn, and for that reason it is sometimes called an *ice giant*.

It's not impossible to see Uranus with the naked eye, but it's very dim. However, unlike the other planets, which were known to naked-eye observers throughout history, Uranus wasn't discovered until the invention of the telescope. The first observer to officially chart it was John Flamsteed (1646–1719) in 1690, and he thought it was a star. This is because it didn't seem to be moving. Sir William Herschel (1738–1822) spotted it in 1781, and he noted in his observation journal that this object seemed to be a comet. This is because through his telescope, Uranus appeared disk-like and fuzzy, not point-like as a star would look. With more observations and some calculations of the object's supposed orbit, it became clear that Herschel's "comet" was really a planet.

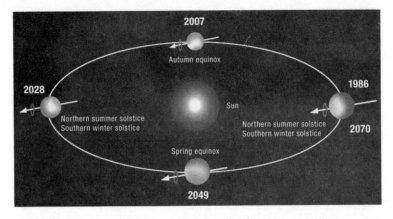

Uranus is tipped on its side, which means that its poles and equatorial regions point toward the Sun during different parts of its orbit.

Uranus Facts

1. Closest point to Sun: 2.7 billion kilometers (1.7 billion miles)
2. Most distant point from Sun: 3.0 billion kilometers (1.86 billion miles)
3. Length of year: 84 Earth years
4. Length of day: 17 hours (based on rotation of interior)
5. Tilt of axis: 97.77 degrees
6. Gravity: 0.89 Earth's gravity

The History of Uranus

The gas giants formed in a region of the solar system where temperatures were cold enough to allow hydrogen and various other compounds to freeze. Some recent research suggests that they actually formed closer to the Sun than where they are now and later migrated out to their present positions.

These worlds were built by planetesimals slamming into one another to build bigger ones, just as other worlds in the solar system did. However, the inner worlds are made mostly of rock, while the outer worlds became mixtures of rock and ice at their cores, covered by layers of liquid metallic hydrogen and helium and smothered in thick atmosphere. Large amounts of methane in the upper atmosphere color Uranus a pale blue-green, and there's a smog layer hiding cloud features from our view.

Uranus wasn't created with its odd axial tilt. Something had to push it over. What could that have been? The best theory is that the infant Uranus was in a collision with an Earth-sized object that was strong enough to tip the planet on its side.

The Moons and Rings of Uranus

Uranus has twenty-seven known moons. Some are named after characters in Shakespeare's plays and from the *Rape of the Lock* by Alexander Pope. Titania and Oberon were discovered in 1787. Ariel and Umbriel were found in 1851, and Miranda was first spotted in 1948. The other moons were found during the *Voyager 2* flyby, and more recently through *Hubble Space Telescope* observations. The five largest moons are Ariel, Umbriel, Titania, Miranda, and Oberon. Each is a small, icy body with many cracks and craters on its surface.

The small moon Miranda looks most unusual. It has a very mottled appearance, with deep canyons splitting its crust. There are also curious-looking, oval-shaped landscapes called *coronae*. In some places, faults (cracks) in the surface crisscross each other. In other places, Miranda is torn apart by the action of cryovolcanism— ice volcanoes spewing softened and slushy ice across the surface. It's likely that Miranda, like the other four large moons, has some kind of geological activity going on inside. This activity causes changes to the surfaces of these moons in the same way that tectonism and volcanism change Earth's surface.

Big Cliff on a Little Moon

When the *Voyager 2* spacecraft swept past Uranus in 1986, it caught a quick image of a most unusual feature on the little moon Miranda: a cliff 5 kilometers (3.1 miles) high. It's surrounded by what looks like cracks in the ground, suggesting that something powerful shook up this moon. This jagged scar on Miranda is named Verona Rupes, and it is the tallest known cliff in the solar system.

Many of Uranus's moons orbit in the same plane as its ring system, which was only discovered in 1977. There are thirteen rings, and the system is fairly dark and hard to spot. The rings formed relatively recently in solar system history—perhaps 600 million years ago. They were probably a result of the collision of some of the planet's moons. The debris was scattered and eventually collected into the flat ring structures we see today. Like Saturn's rings, which are at most a few kilometers thick, the Uranian rings are fairly thin. They stretch out as far as 98,000 kilometers (nearly 61,000 miles) from the planet and are made up of small water ice chunks and dust particles.

Exploring Uranus

Until the *Voyager 2* spacecraft encountered Uranus in 1986 (the only spacecraft to do so), exploration of this distant ice giant was done using ground-based telescopes. As *Voyager* passed 81,500 kilometers (50,644 miles) above the cloud tops, the probe snapped images, studied the gases in the planet's atmosphere, charted the peculiar magnetic field, and captured views of five of its largest moons.

The *Hubble Space Telescope* has also studied Uranus from its vantage point in Earth orbit. The telescope's long-term mission is to provide frequent monitoring of activity in the Uranian atmosphere, track seasonal changes, and look for more auroral displays (such as the northern and southern lights here on Earth) driven by interactions between Uranus's magnetic field and streams of charged particles from storms that originate on the Sun and are carried along on the solar wind.

There are no further missions to Uranus underway, although some are on the drawing board. One idea is to mount a joint mission by NASA and the European Space Agency, which would launch in 2022. Another proposal would send an orbiter to Uranus, where it would spend two years studying this distant and still mysterious system.

How Uranus Got Its Name

Uranus (pronounced *YOU-ruh-nuss*) was not this planet's first name. Sir William Herschel wanted to call the planet Georgium Sidus, after King George III, who was his patron. That didn't sit well with observers outside of Great Britain, and so after some discussion, the seventh planet from the Sun was named Uranus. It was a popular choice, and not long after the planet's discovery, the German chemist Martin Klaproth discovered and named a new radioactive element after it: uranium.

NEPTUNE

The Outermost Planet

The last major planet in the solar system, Neptune is another world of superlatives. It's the farthest planet from the Sun, at an average distance of 4.5 billion kilometers (2.8 billion miles). It takes 165 Earth years to circle the Sun once. At 24,764 kilometers (15,338 miles) across, it's the smallest of the gas giants. It's also the coldest world of all the planets, with temperatures dipping down to –221°C (–365.8°F) at the top of its atmosphere. Neptune also has the strongest winds. They blow up to 2,100 kilometers (1,304 miles) per hour. Those winds, in combination with the rotation of the atmosphere and planet, the almost Earth-like tilt of its axis, plus an unknown source of internal heat, give Neptune distinct seasons and drive some very fierce storms.

Neptune has a layered interior structure somewhat similar to the other gas giants, but with a few differences. There's a small rocky core, covered with a mantle made of a mixture of water, methane, and ammonia ice. Above those are a lower atmosphere consisting of hydrogen, helium, and methane gases and an upper atmosphere with clouds that contain ice particles. The methane gases in the top of the atmosphere give Neptune its characteristic deep blue color. Since Neptune has a higher ice content than Jupiter or Saturn, it's often referred to as an *ice giant*.

Like the other planets of the solar system, Neptune formed in the same cloud of gas and dust as the Sun some 4.5 billion years ago. It may have formed closer to the Sun, along with the other planets, and then migrated out to its present position.

Neptune Facts

1. Closest point to Sun: 4.4 billion kilometers (2.7 billion miles)
2. Most distant point from Sun: 4.5 billion kilometers (2.8 billion miles)
3. Length of year: 164.8 Earth years
4. Length of day: 16.1 hours (based on rotation of interior)
5. Tilt of axis: 28.32 degrees
6. Gravity: 1.12 Earth's gravity

Neptune's Moons and Rings

There are thirteen known moons orbiting Neptune. They probably didn't form with Neptune, but were captured later on by the planet's gravitational pull. The largest moon, Triton, was discovered in 1846 and was imaged in great detail by the *Voyager 2* spacecraft in 1986. Triton orbits its planet in what's called a retrograde orbit. That means it orbits against Neptune's rotation. This alone is a good clue that Triton didn't form with the planet but was captured later on. It's 2,700 kilometers (nearly 1,700 miles) across, making it one of the largest moons in the solar system.

Like its sibling gas giants, Neptune has a ring system, although it's not nearly as massive as Saturn's. Very little is known about the particles that make up the rings, but scientists suspect that they may be ice particles coated with some kind of complex organic material. Oddly enough, the rings seem to be disappearing, and the reasons why are still being debated.

Triton: The Active Moon

Although Triton had been observed from Earth for years, nobody expected to find what *Voyager 2* found when it swept past in 1989. This frozen moon has a mottled surface made of nitrogen, water, and methane ices. A large part of the surface is covered with *cantaloupe terrain*, so called because it looks like the skin of that fruit. There's also a rock-metal core hidden deep inside. Most fascinating of all is that some kind of internal activity drives nitrogen geysers up into Triton's thin atmosphere. They show up as dark plumes that rise some 8 kilometers (about 5 miles) above the frosty surface.

Who Discovered Neptune?

The tale of Neptune's discovery is a triumph of mathematical prediction by a man named Urbain Le Verrier (1811–1877). Actually, Neptune was inadvertently observed by such a luminary as Galileo Galilei, who thought he'd found a star. Others spotted this dim object (which can't be seen with the naked eye), but failed to recognize it as a planet. Fast-forward to the 1840s, when observers were trying to determine what was perturbing the orbit of Uranus. Several predicted that the gravitational pull of another planet was tugging on Uranus. People set to work doing mathematical calculations, and in 1846, Le Verrier presented his work showing where an outer planet might be found.

Another astronomer, John Couch Adams, had been working on similar calculations and also predicted the position of a hidden planet, although not as accurately as Le Verrier had done. On September 24, 1846, astronomers in Berlin, using Le Verrier's predictions, found Neptune. After a great deal of back-and-forth between France and England, both Adams and Le Verrier were given credit for the discovery. In recent years, new evidence has been found to support Le Verrier's claim to Neptune's discovery.

Exploring Neptune

Only one spacecraft has ever visited Neptune. On August 25, 1989, the *Voyager 2* mission flew by the planet. It sent back a collection of high-resolution images of Neptune and Triton and probed the strength of the planet's magnetic field. Oddly enough, that magnetic field is tilted away from the rotational axis of the planet. *Voyager 2* discovered six moons and a ring that hadn't been observed from Earth. While the spacecraft was in the system, it also captured images of several storms that astronomers nicknamed the Great Dark Spot, Scooter, and Dark Spot 2.

Most of what we know about Neptune these days comes from observations made using ground-based telescopes and the *Hubble Space Telescope*. There are no active plans for new missions to Neptune, although there is ongoing research into new generations of long-duration robotic probes that could be sent to the outer planets in the distant future.

The Great Dark Spot

One of the great finds of the *Voyager 2* mission was the discovery of a giant anti-cyclonic storm raging through the planet's upper cloud decks. It looked dark, so it was immediately dubbed "The Great Dark Spot." It lasted for several years, disappearing before it could be imaged by the *Hubble Space Telescope* (which had not yet been launched when the *Voyager 2* spacecraft got to Neptune). Other storms have come and gone, and in recent years a new "Great Dark Spot" has emerged in the northern hemisphere.

PLUTO

The Dwarf Planet

Many people were shocked in 2006 when the International Astronomical Union announced that Pluto was no longer a planet. Instead, it was henceforth to be known as a dwarf planet, along with other outer solar system worlds such as Quaoar, Makemake, Eris, and Sedna. This didn't change the fact that Pluto has been the "King of the Kuiper Belt" for decades, but in the eyes of the public, the demotion of Pluto was sad news. The media got involved and there were dueling scientists on TV talk shows, all over an attempt to define planets more rigorously than in the past.

Pluto is still the same world it was before the announcement. The change in status from planet to dwarf planet didn't affect its physical parameters one bit. But it does allow better classification of the worlds of the outer solar system, and that's important. First, Pluto lies some 5 billion kilometers (3.1 billion miles) from the Sun, and it takes 248 years to make one orbit around the Sun. Pluto was discovered in 1930 by Clyde Tombaugh (1906–1997), and it hasn't even made one complete orbit since its discovery.

Pluto Facts

1. Closest point to Sun: 4.4 billion kilometers (2.7 billion miles)
2. Most distant point from Sun: 7.3 billion kilometers (4.5 billion miles)
3. Length of year: 247.8 Earth years
4. Length of day: 6.39 Earth days (retrograde)

5. Tilt of axis: 123 degrees

6. Gravity: 0.9 Earth's gravity

Pluto orbits the Sun in a region of the solar system called the *Kuiper Belt*. It stretches out from the orbit of Neptune and probably contains many worlds the size of Pluto (or even larger). Like many of those other objects, Pluto is made of rock and ice, with a surface made almost entirely of nitrogen ice mixed with small amounts of carbon dioxide and methane. Pluto probably has a rocky core surrounded by a water ice mantle. There may be some kind of heating going on inside, since patchy surface features seem to indicate that some kind of cryovolcanic activity is forcing icy material out from beneath the surface.

Pluto's Moons

From its frozen icy surface, the Sun would look very much like a large, bright star. If you could stand on Pluto, you'd see its largest moon Charon off in the distance. There are also four other moons, all recently discovered in observations made by the *Hubble Space Telescope*. Charon is actually classified as a dwarf double planet with Pluto. They're locked together in what's called a tidal resonance, which means that they each present the same face to each other as they orbit around their common center of gravity.

The moons Nix and Hydra were discovered in 2005. They have bright and dark patches on their surfaces, which suggests the existence of ice deposits. Two other very small moons temporarily named P4 and P5, have been found. One was discovered as part of a search for hazards in the Pluto system that might affect the upcoming *New Horizons* mission to the outer solar system.

What Are Dwarf Planets?

When the International Astronomical Union debated the definition of *planet* and came up with a set of criteria that excluded Pluto, Sedna, Eris, and other newly discovered objects in the Kuiper Belt, the members also devised a new definition for those worlds. They are dwarf planets—worlds that orbit the Sun, and are rounded by their self-gravity, but have not yet cleared their orbits of other planetesimals. They're also not satellites of other worlds. Currently there are five recognized dwarf planets:

1. Pluto (with its companion Charon)
2. Ceres
3. Haumea
4. Makemake
5. Eris

These last three were discovered in 2004 and 2005, and their discovery played a large role in the discussion about redefining planets and dwarf planets. The issues are still being debated, and there may be further clarifications coming.

Where Did Pluto Come From?

Pluto has an eccentric orbit, which means that it follows a very elliptical path around the Sun. Other planets' orbits are more circular while Pluto's path is highly inclined. Most of the other worlds orbit near the plane of the solar system, but Pluto's orbit takes it above and below the plane. That eccentricity and high inclination tell an interesting tale of Pluto's origins. First, it's probably a planetesimal left over from the formation of the solar system some 4.5 billion years ago. It never coalesced with other pieces to form a planet. At one time, it was thought that perhaps Pluto had been captured from outside the solar system, since that would explain its strange orbital characteristics. However, early in solar system history, the gas giants likely formed much closer to the Sun. At some point, they migrated to their current positions in the outer solar system. Neptune's migration likely swept Pluto and other similarly sized Kuiper Belt objects into their current positions and orbits.

Pluto's Discovery

Pluto is tough to spot. It can't be seen without a good-sized telescope and a lot of patience. In the late nineteenth century, astronomers speculated about whether or not there even *was* another planet out beyond Neptune, the latter having been discovered because of its effect on the orbit of Uranus. Further observations of Uranus showed that there were still some effects not accounted for by Neptune's presence. The idea of a Planet X spurred Percival Lowell (1855–1916), who founded Lowell Observatory in Flagstaff, Arizona, to mount a search for a ninth planet. As it turns out, Lowell and others *did* photograph Pluto, but they didn't know what they had. It took a determined search by a young man named Clyde Tombaugh to find this elusive object. He made photographic plates of the sky where Pluto was predicted to be and then compared them to find an object that seemed to move. When he found his quarry on February 18, 1930, the news galvanized the world. It was the first planet discovered in the new century, and it was found by an American. At the suggestion of an English schoolgirl named Venetia Burney, Tombaugh named his discovery Pluto and noted that the first two letters were also a salute to Percival Lowell, who had died some years earlier. Later, it turned out that the extra perturbations in Uranus's and Neptune's orbits were the result of a mathematical error, and not due to Pluto's influence.

Exploring the Outer Frontier

The exploration of Pluto has, for many years, come through ground-based observatories and orbiting instruments such as the *Hubble Space Telescope*. That will change in 2015 when the *New Horizons* mission will fly past Pluto and its moons and give the first up-close views of this distant world. After that, the probe will move farther out through the Kuiper Belt, examining other objects in this new frontier of the solar system.

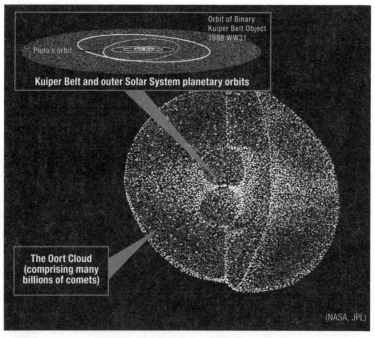

Orbit of Binary
Kuiper Belt Object
1998 WW31

Pluto's orbit

Kuiper Belt and outer Solar System planetary orbits

**The Oort Cloud
(comprising many
billions of comets)**

(NASA, JPL)

The Kuiper Belt is a disk-shaped region in the outer solar system that stretches out beyond Neptune. It lies inside the Oort Cloud, which is shaped more like a sphere. The Kuiper Belt is filled with small icy worlds and comet nuclei. The Oort Cloud also contains countless cometary nuclei.

COMETS

From Ancient Harbingers to Modern-Day Observing Targets

For much of human history, people were wary of comets. Early sky gazers had little understanding of what celestial objects really were. To them, comets weren't like stars or planets. They seemed to appear and disappear without warning, and they changed shape over time. Some observers thought they were harbingers of bad fortune and bringers of doom, warning of catastrophic events about to happen.

Today, we know that comets are solar system objects in orbit around the Sun. They consist of a central core, called a *nucleus*—which is a combination of ices and dust. As a comet gets close to the Sun, those ices begin to vaporize—the technical term is *sublimation*. This is similar to what dry ice does if you expose it to air: It goes directly from an ice to a vapor. The materials in the comet form a cloud around the nucleus called a *coma* and also stream away from the comet in a pair of tails. The *dust tail* is made of dust from the comet, while the *plasma tail* consists of ionized (heated) gases that glow as they encounter the solar wind.

Interestingly, comets are known to leave behind streams of particles, which are eventually spread out along the path of the comet's orbit. When Earth encounters one of those streams, the material gets caught up in our atmosphere. As particles from the stream fall to the surface, they vaporize in the atmosphere, creating meteors.

The Oort Cloud

There is a vast collection of icy objects at the outer limits of our solar system called the Oort Cloud, for the Dutch astronomer Jan Oort (1900–1992), who first suggested the idea. This spherical cloud may stretch out about a light-year from the Sun. It's not clear exactly how many comet nuclei exist out there or how the cloud formed. One idea is that it may have been assembled when icy materials from the inner regions of the solar system migrated out with the giant planets as they formed and moved into their current positions.

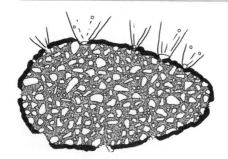

Comets have been called "dirty snowballs" and are thought to be a mixture of ices and rock and dust particles. As the comet gets closer to the Sun, the ices sublimate and rush out from the interior of the comet as jets of particles.

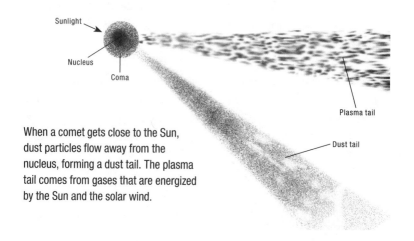

Sunlight

Nucleus

Coma

Plasma tail

Dust tail

When a comet gets close to the Sun, dust particles flow away from the nucleus, forming a dust tail. The plasma tail comes from gases that are energized by the Sun and the solar wind.

Our Changing View of Comets

The earliest scientific thinking about comets began with the Greek philosopher Aristotle, who suggested they were the result of events happening in the upper atmosphere of our planet. There was a problem with that idea, however. If comets were very close to us, then they would look as if they were moving quickly through space. Comets change their positions very slowly, taking weeks or months to cross the sky. So what could they be, and where were they?

Tycho Brahe (1546–1601), a Danish astronomer, took immense interest in a bright comet that appeared in 1577. He and others carefully measured its position over several months and determined that it was well beyond Earth. A century later, the appearance of another bright comet allowed observers to determine its orbit by applying Kepler's ideas about how objects moved in orbit around the Sun. Later on, Sir Isaac Newton (1642–1726) also showed that comets do move in such orbits.

Following the Dusty Tail

The tails of comets are sometimes depicted as if they stream out along the direction of the comet's travel. This is misleading. Comet dust tails point away from the Sun, while the plasma tails are carried along by the solar wind. So, sometimes a comet can appear to have two tails pointing in very different directions.

The next bright comet to change our view was the great one: Comet Halley. Astronomer Edmond Halley (1656–1742) studied apparitions of comets noted in historical records and recognized that one particular comet kept showing up every seventy-six years. He used that information to predict its next appearance in 1758, although the comet actually showed up in 1759 due to changes in its orbit introduced by the gravitational pull of the outer planets. Comet Halley last appeared in our skies in 1985–86, and will make its next visit in the year 2061.

Halley's Comet Makes News Throughout History

Working backward from Halley's calculations, historians realized that Halley's Comet must have been the same one that appeared in the skies over Europe in the year 1066 and was depicted on the Bayeaux Tapestry, which told the story of William the Conquerer's successful invasion of England. Babylonian cuneiform clay tablets record apparitions of the comet as early as 164 B.C.E.

The twentieth century brought many advances in comet science, including Fred Whipple's (1906–2004) idea that comets are essentially dirty snowballs in space. This is now the commonly accepted view. But where do comets come from, and how do they get flung into orbits around the Sun?

Comet Names

Comet Halley gets its name from Sir Edmond Halley. For a long time comets were simply named after their discoverers. Today their names also include the year of discovery, the half-month of the discovery, and other letters that indicate additional information about the comet. Under this system, Comet Halley is now known as 1P/Halley. This means it's the first periodic comet discovered, credited to Halley. Another comet, discovered in 2012 by observers using the International Scientific Optical Network, is called C2012 S1 (ISON).

Treasures Across Time

Today we know that comets are relics of the materials that existed when the solar system began forming some 4.5 billion years ago. This makes them treasuries of solar system history. In recent years, several missions—including *ICE*, *Vega 1* and *2*, *Suisei*, *Giotto*, *Deep Space 1*, *Deep Impact*, and *Stardust*—have actually visited comets or captured samples of comet material for study. The European Space Agency *Rosetta* mission is aimed at a rendezvous with comet 67P/Churyumov-Gerasimenko and will land a probe on its surface.

During the formation of the solar system, the protosolar nebula was probably rich in icy materials throughout the cloud. Some ices were destroyed by the heat of the Sun or accreted into the planets. The rest probably migrated to the outer solar system, to create an icy shell of cometary nuclei called the Oort Cloud. There are also cometary nuclei in a region of the solar system called the Kuiper Belt. It extends out from the orbit of Neptune and contains many smaller frozen worlds and cometary nuclei.

A Headlong Rush to Glory Around the Sun

Comets travel long distances in their orbits—some from the most distant reaches of the solar system. It's possible that some kind of gravitational interaction nudges a nucleus out of its comfortable orbit in the Oort Cloud. Perhaps a chunk of ice and dust gets moved by a close call with a nearby neighbor in the Kuiper Belt or by a brush with Neptune. That's enough to change the object's orbit toward the Sun. Gravitational nudges from other worlds along its path subtly alter the comet's path. From there, it's only a matter of time until the comet gets close to the Sun, grows a tail, and shows us the familiar shape that has fascinated observers throughout history.

METEORS AND METEORITES

What's Behind the Flashes in the Sky?

Ever seen a shooting star? If so, you've watched the evaporation of a bit of solar system history. Technically, what you saw was a piece of solar system debris vaporizing in Earth's atmosphere and leaving behind a visible path called a meteor. If any piece of it survived the trip and fell to the ground, the remaining rocky bits would be called *meteorites*. Some of these pieces of the solar system may have existed long before the Sun and planets formed. Others might have been part of an asteroid or pieces of the Moon or Mars.

Think You Have a Meteorite?

People often run across rocks they think are meteorites. A true meteorite will be very dense and feel quite heavy. It's usually magnetic and may have small inclusions called *chondrules* embedded in it. Most meteorites have what's called a *fusion crust*, which is a black, ash-like, darkened exterior. The surface also has smoothed depressions and cavities that look like thumbprints. These are called *regmaglypts*. Finally, most are made of a mixture of iron and nickel. If you find something you think might be a rock from space, take it to a local observatory or a college geology department and ask someone to look at it. Many Earth rocks and iron waste products from industry get mistaken for meteorites, so it's always best to ask an expert if you run across a strange-looking rock out in the field.

The Bombardment of Earth

Each day and night, our planet is showered with more than a hundred tons of material from "out there." Most of the pieces are very small—perhaps the size of dust particles. These don't last very long during the trip through Earth's thick atmosphere. Friction heats them up and they simply evaporate. Bigger pieces survive the trip to the surface where they plunge into the ocean or splash into lakes, such as the chunk of rock that came to Earth in Russia on February 15, 2013. Some of the meteorites end up on land, where they eventually get picked up by collectors. But what about really big pieces of solar system debris? Can they hurt us?

To answer that, let's look back at solar system history. Early in Earth's formation, it was constantly being bombarded. Oddly enough, though, if it hadn't gone through that pummeling, our planet wouldn't exist. Bombardment and accretion built our planet. Even after it was fully formed, debris kept raining down on the planet, adding more material and scarring the young landscape with craters. Bombardment still occurs, although the pieces that fall to Earth today are fairly small and usually cause little damage.

Cosmic Collisions

Collisions still take place regularly in the Asteroid Belt, and many asteroids could be collisional debris that has reassembled into floating rubble piles.

There is still a large amount of solar system debris left over, even after several billion years of bombardment. Some are large enough—a kilometer (0.6 mile) across—to cause big problems if they landed on Earth. These don't hit us very often—maybe once every million years or so. When they do, they can cause catastrophic damage. So planetary scientists are on the lookout for objects around this size that are in Earth-crossing orbits. We hear about them from time to time, and when they're discovered, observers spend a lot of time calculating accurate orbits so they can predict whether or not these objects pose a danger to Earth. The Catalina Sky Survey in Arizona is one of many "early warning" observation projects keeping a lookout for potentially hazardous asteroids that could collide with Earth.

Where Do Meteors Come From?

Most of the meteors you see in the night sky are little specks of dust or tiny pieces of rock. Many of them are left behind as comets travel around the Sun, shedding dust and ice as they go. If Earth's orbit intersects those cometary streams, we experience a meteor shower as meteoroids rain down through the atmosphere.

Meteor Showers

There are eight major meteor showers each year. To get the best view, plan to stay up late at night or observe early in the morning. You might see dozens of meteors per hour, including some *bolides*, which are the trails of large pieces of debris that move through our atmosphere. Meteor showers are named after the constellation they seem to radiate from in the sky. The best-known are:

- January: Quadrantids
- April: Lyrids
- May: Eta Aquarids
- June: Arietids and Bootids
- July: Southern Delta Aquarids
- August: Perseids
- October: Orionids
- November: Leonids
- December: Geminids

Each of these, except the Geminids, is caused by Earth moving through a stream of comet debris. The Geminids come from a stream of debris from the Asteroid 3200 Phaethon, which is probably a dead comet.

Some meteors come from the debris left over from collisions of asteroids. The shattered remains scatter through space, and eventually they stray across Earth's orbit and encounter our atmosphere. Other meteors are chunks of debris that are shot into space when other objects collide with the Moon. The collision blasts lunar rocks out from the surface, and eventually they intersect Earth's orbit and come plunging down through our atmosphere.

The rarest types of meteorites found on Earth come from Mars. They have experienced a long journey—sometimes millions of years. How do rocks from Mars get to Earth? In a way similar to how Moon rocks get to our planet. Mars has been bombarded throughout its history, which you can tell by looking at its many impact craters. During an impact, surface rocks from Mars can be ejected into space. They fall into orbits that eventually take them close to Earth's orbit, and if they stray too close, they are swept up by Earth's gravity and fall to the surface.

Meteorites are treasure troves of information about the conditions under which they formed in the early solar nebula, and about the comets, asteroids, Moon, or part of the Martian surface where they originated. This is why astronomers are so interested in studying meteorites that hit the ground. They're precious samples from the depths of the solar system.

Meteorite Types

There are three basic types of meteorites. The irons come from the core of an asteroid or planetoid and are quite heavy and massive. The stony types come from the crust of a planet or asteroid. They often contain little rounded grains called *chondrules*, and these chondritic objects are at least as old as the solar system itself. Finally, there are the stony-irons. They are somewhat rare and come from a region between the core and the mantle layer of an asteroid or early planet.

ASTEROIDS

Solar System Debris

Millions of bits and pieces of solar system debris are orbiting among the planets. They're called *asteroids,* and they represent a class of objects called planetesimals—jagged-looking little bodies that existed in the early solar nebula as the planets were forming. The asteroids that exist today are the ones that didn't get accreted into larger worlds, or they were part of early versions of the planets that got smashed apart during collisions. Most asteroids orbit between Jupiter and the Sun and are made primarily of rocky materials.

Asteroids are classified by the spectra of their reflected light. The chemical elements in asteroids can be determined by looking at that light through an instrument called a *spectroscope.* It splits the light into its component wavelengths. Each chemical element that exists on the surface of the asteroid has a characteristic set of lines in the resulting spectrum.

Naming an Asteroid

Whenever a new asteroid is found, it is given a temporary designation that consists of the year it was discovered and a code that tells the half-month of the discovery. Eventually the asteroid gets a number, and it may get a name. The discoverer has the right to bestow a name, and so we have asteroids such as 433 Eros (named after a mythical god) or 3503 Brandt (named for an astronomer). There are also asteroids named for writers, musicians (each of the Beatles has an asteroid), and many others.

Asteroids are grouped into three main types:

1. **C-type asteroids** are carbonaceous. That means they're made mostly of carbon compounds mixed with hydrates (minerals that contain water). Scientists think these existed in the young solar nebula as the Sun began to form.
2. **S-type asteroids** have a composition that is close to stony meteorites. That means they're made mostly of iron and a mixture called a magnesium silicate (similar to rocks in Earth's crust).
3. **M-type asteroids** are mostly made of a nickel-iron mix and could be from the metallic cores of larger objects broken apart by impacts.

The First Asteroid Discovered

The largest asteroid in the solar system is called 1 Ceres. It's also a dwarf planet. Ceres was discovered in 1801 and its discoverer, Giuseppe Piazzi (1746–1826), considered it a planet. Ceres is a rounded world and doesn't look too much like the rest of the asteroids. It has a light-colored surface that's probably covered with water ice and clays. The *Dawn* mission that visited the asteroid 4 Vesta in 2012 is now on its way to study Ceres.

Asteroid Groups and Families

Most of the asteroids in the solar system are part of the Asteroid Belt, a collection of several million asteroids orbiting between Mars and Jupiter. They occupy a region of space where a planet could have formed during the birth of the solar system, but the gravitational influence of Jupiter may have kept the pieces from coming together.

Other asteroids are grouped into families that have similar paths through space, implying that they may have once been part of a single larger body that was broken and scattered into orbit by collisions. There are around thirty different asteroid families, and many of those lie in the main Asteroid Belt. Others orbit at different distances, some as far out as the Kuiper Belt. One well-known group orbits in the vicinity of the large asteroid 4 Vesta. These are known as *vestoids*. They got scattered into orbit when something hit the asteroid Vesta in the distant past, creating a huge crater.

Other groups of asteroids orbit in the same region of the solar system as another planet or even a moon. Most so-called "Trojan Asteroids" are found in the same orbit as Jupiter. Others can be found in the orbit of Earth, Mars, and Neptune. The name "trojan" comes from an orbital configuration where an object such as an asteroid shares an orbit with a planet or a larger moon. They don't collide because they're in stable points in the orbit called "trojan points." These are also known as "Lagrangian points."

Near-Earth Asteroids

A good number of asteroids in the inner solar system have orbits that cross Earth's. They're called Earth-crossers, and it is possible that their orbits can intersect with ours from time to time, bringing an asteroid dangerously close to our planet. Some asteroids come close to Earth but don't always cross our orbital path. They're called Near-Earth Asteroids (NEAs for short). At least 10,000 of these close neighbors have been found, and up to 1,000 of those are larger than a kilometer (0.6 mile) across. These asteroids are grouped into several categories according to the danger they pose. Those that are the greatest threat are called Potentially Hazardous Asteroids (PHAs), and once they are discovered, they're tracked very carefully. These are particularly sensitive to gravitational tugs from Earth. During a close flyby of Earth, a PHA might miss the planet, but it could pick up a gravitational kick that could re-aim it toward Earth on a future orbital pass. This is why astronomers spend a lot of time accurately calculating the orbits of PHAs.

Visiting an Asteroid

Asteroids are studied through both ground-based and space-based telescopes, and also by spacecraft. The *Galileo* probe was the first to visit an asteroid when it swept past asteroid 4660 Nereus on its way to Jupiter. After that, the *Cassini* mission to Saturn took data as it flew through the Asteroid Belt in 1997. Following that was the *Near-Earth Asteroid Rendezvous* mission, which visited 433 Eros in 2001. *Deep Space 1* went to asteroid Braille; *Stardust* gathered information about asteroid AnneFrank; the Japanese *Hayabusa* probe went to 25143 Itokawa; and the European Space Agency's *Rosetta* craft went to asteroids Steins and Lutetia.

THE STARS

From Ancient Sky Lamps to Astrophysical Wonders

Since the beginning of time, the starry night has fascinated us. Think about the first intelligent beings on our planet and what the stars must have meant to them. They saw thousands of bright, glittering points of light. It must have been a breathtaking view since they had no light pollution to wash out the sky. They didn't know *what* those things were up there, but that didn't stop them from making up stories about them and transmitting those tales to later generations. Today, we twenty-first-century stargazers have the same reaction our ancestors did to a starry sky: amazement and wonder. When we look at the night sky, we know that we're seeing stars of all ages and states of evolution, from the very young to the very old. In addition, we have at our fingertips a vast amount of knowledge about stars. It comprises the science of stellar physics.

Star Names

People often want to know if they can name stars. Some companies sell star names as "novelties," but those are not used by astronomers. The official star names astronomers employ come from a variety of cultural and scientific sources. For example, Arcturus and Betelgeuse are from Greek and Arabic words, respectively. The bright star Sirius (which is also a Greek name) is also known as α Canis Majoris (Latin for "big dog") and written as α CMa. This is because it's the brightest star in the constellation Canis Major. The "α" indicates it is Alpha, the brightest star. Beta, or β, is the second-brightest star, and so on. Sirius is also called HIP 32349, from the *Hipparcos* satellite sky survey. If you see something labeled *NGC*, it's from the New General

Catalog, a listing of deep-sky objects such as star clusters, nebulae, and galaxies. Charles Messier (1730–1817), a French astronomer, also compiled a list of faint objects. The Orion Nebula, which is labeled NGC 1976, is Messier number 42 or M42.

Peeking Inside a Star

There are at least a few hundred billion stars in our Milky Way Galaxy. Each one is a luminous sphere of superheated matter called plasma—gases that are energized and heated—held together by self-gravity. Stars fuse elements in their cores to make other elements. That process, called *nuclear fusion,* generates heat and light. The action of fusion and the self-gravity of the star combine to make a star a stable sphere of superheated material. In the chapter about the Sun, you can dive into our own star to see this process at work.

Stars fuse hydrogen into helium through much of their lives. As they age, they run out of hydrogen, and some begin fusing the helium into carbon, and so on up to iron in the most massive stars. This process of creating new elements is called *stellar nucleosynthesis.* When a star dies, the material it created is distributed into space and recycled into new generations of stars, planets, and life. The astronomer Carl Sagan once pointed out that "we are star stuff." The iron in our blood, the oxygen we breathe, the calcium in our bones, the carbon in the molecules of our cells, all came from the insides of long-gone stars. To find out more about how stars are born, see the Star Birth chapter, and to learn how they die, see the Star Death chapter.

Types of Stars

Astronomers classify stars by their color, which reveals the temperature at the star's photosphere (the visible "surface"). They pass starlight through a spectroscope, an instrument that splits light into its component wavelengths (creating a spectrum). Each element that exists in the star leaves distinctive fingerprints in that spectrum in the form of dark dropouts called *absorption lines*. Those lines show that the temperature of the star is just right for that element to exist there.

Stellar spectra are incredible tools. They tell us:

- How fast a star is rotating
- What chemical elements a star contains (its metallicity)
- A star's speed through space
- The strength of a star's magnetic field
- The approximate age of a star

There is a standard set of star types called *spectral classes*, and most stars fit into them quite well. The classes are O, B, A, F, G, K, and M, with O-type stars being the hottest, densest hypergiants and K and M stars being the coolest red dwarfs and supergiants. Some classifications contain even cooler-temperature objects that might be brown dwarfs (objects too hot to be planets and too cool to be stars) or even rogue planets. They are labeled L, T, and Y, and some of them are known to be cooler than planets. The types also include Roman numerals and letters to indicate other characteristics of a star. Our Sun is a G2V-type dwarf star, which means that it's about average in terms of temperature. The bright star Betelgeuse, in the constellation Orion, is classified as M2Iab. That means it's an M-class intermediate luminous supergiant star.

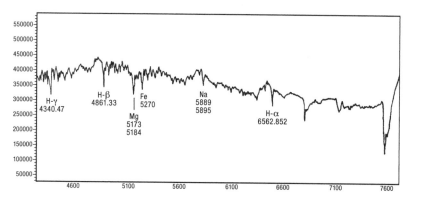

This spectrum of the Sun shows which elements are abundant in its atmosphere because those elements absorb light and cause the dips seen here. They show there is plenty of hydrogen, plus sodium (Na), iron (Fe), and magnesium (Mg).

How Long Do Stars Exist?

Stars' lifespans can be measured in millions or billions of years. Most stars we see in the sky now are between 1 and 10 billion years old. The more massive a star is, the shorter its lifespan. The Sun is an intermediate-mass star. It formed about 4.5 billion years ago and will last for another 5 or so billion years before it ends up as a white dwarf. A very massive star—one that is many times the mass of the Sun—might live only a few million years before exploding as a supernova. Stars with very low masses exist seemingly forever because it doesn't take much to keep them going. These so-called *red dwarfs* can exist many billions of years before they finally cool into cinders. So far as we know, no red dwarf has ever died out since the beginning of the universe!

The Evolution of a Star

Astronomers like to make graphs to help them understand how a star will evolve over time. The most famous one they use is the Hertzsprung-Russell Diagram. It's a plot of stellar temperature versus luminosity. Most stars fall along a narrow, winding band on the plot called the Main Sequence and stay on it as long as they are fusing hydrogen in their cores. A star's mass determines how long it will stay on the Main Sequence. Once it stops fusing hydrogen, its color and brightness change. Eventually the star gets plotted off the Main Sequence and onto another part of the diagram. Stars smaller than about a quarter of the Sun's mass evolve to become white dwarfs. More massive stars, including the Sun, swell up to become red giants and eventually white dwarfs. The most massive stars become red supergiants.

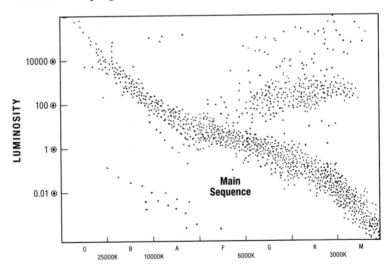

The Hertzsprung-Russell Diagram arranges stars according to their temperatures and luminosities (brightnesses).

Star Bright, Star Light!

Stars, planets, nebulae, and galaxies have different brightnesses. The word *magnitude* refers to how bright they look. *Apparent magnitude* is the brightness of a star as it appears from Earth. *Absolute magnitude* indicates how bright objects would appear if they were located ten parsecs away. Putting all stars at a standard distance is a good way to tell the true brightness of an object compared to others. The bigger the magnitude number, the dimmer the object. The smaller the number (and this includes negative numbers), the brighter the celestial object. The faintest objects seen by the *Hubble Space Telescope* are about magnitude 30. Our eyes can see only to about magnitude 6. Beyond that, we need to use binoculars or a telescope to see dimmer objects. On the other hand, the Sun is very bright and listed as magnitude –27.

Stars in the Wild

Most stars are in multiple-star systems such as double stars, triples, and clusters. Loners such as our Sun probably formed in clusters but scattered after their births. The largest such groupings are *open star clusters*, which contain up to a few thousand stars, and *globular clusters*, which are spherical collections of up to hundreds of thousands of stars. Individual stars and star clusters are grouped into galaxies—collections of hundreds of billions of stars. These are the true stellar cities of the cosmos.

STAR CLUSTERS

Batches of Stars

Many stars in the Milky Way Galaxy spend at least part of their lives in clusters. There are two types—*open* and *globular*. Open clusters usually have up to a thousand or so stars gathered into an irregularly shaped collection. They are often found in the plane of the galaxy, which is where they form. Most of the stars in these clusters are less than 10 billion years old, and some still lie embedded in what's left of their birth clouds. Our Sun was created in an open cluster that formed about 4.5 billion years ago. It has since moved away from its stellar siblings and now travels the galaxy alone.

Globular clusters are collections of hundreds of thousands of old stars. The gravitational influence of all those stars binds the cluster together into a spherical, globular shape. Globulars swarm around the central region of the galaxy, called the halo. The Milky Way Galaxy has about 160 of these tightly packed clusters, but other galaxies have many more.

Clusters are scientifically interesting because all their stars formed around the same time and generally have similar characteristics. For example, if the cloud in which they formed was rich in certain kinds of elements, then the stars from that cloud will contain higher amounts of those materials. If the cloud was metal-poor (that is, it had a lot of hydrogen and helium but very little of other elements), then the stars that form will reflect that metallicity.

Their similarity makes cluster stars good targets for the study of stellar evolution (how stars age and die). Very young clusters interact with the remains of the gas and dust cloud from which they formed. Open clusters are generally found in spiral galaxies such as the Milky Way and irregular-type galaxies such as the Large and Small Magellanic Clouds, which are two of our galaxy's closest neighbors. Globulars roam around the halo and probably formed about the same time as the galaxy did. Understanding how all types of clusters form in our galaxy gives astronomers good insights into how the process happens in other galaxies as well.

Cluster Formation

As with the births of all stars, the formation of a cluster begins in a giant cloud of gas and dust, often referred to as a molecular cloud. Open clusters are born in molecular clouds that lie in the plane of the Milky Way. The formation of globular clusters took place early in our galaxy's history, and that process is still being heavily researched.

The Sun's Siblings

If the Sun was born in a cluster, what happened to its crèche-mates? Nobody's quite sure, but people have been looking for them. To locate stars from the Sun's birth cluster, astronomers study the spectrum of light shining from nearby stars. They compare what they find with the spectrum of the Sun. If they spot similar lines indicating the presence of certain elements in the spectra, and if the stars are moving through the galaxy in roughly the same direction as the Sun, then they could be related. There are two likely solar siblings within 325 light-years of Earth, and there may be more.

A cluster begins to form when some event triggers motion and turbulence in the birth cloud. For an open cluster, it could be a supernova explosion or a fast-moving wind ejecting material from an aging star in the near neighborhood. For a globular cluster, a galaxy collision could be one kind of trigger event. Whatever happens, it sends fast-moving material and shock waves through the birth cloud and starts the process of star birth.

Once formation is complete, the cluster stars continue to evolve. If they are not strongly bound together by gravity, after about 100 million years they start to go their separate ways. Even though members of the cluster may get separated by large distances, they all tend to move through space in the same direction and at about the same speed. Sometimes interactions in the cluster will "kick" some stars out into space, sending them on radically different trajectories into the galaxy. Eventually, these stellar associations dissipate into what's called a moving group, before they finally scatter to become part of the larger stellar population in the galaxy.

Did You Know?

Globular clusters are among the oldest star collections in the universe. They form early in a galaxy's history. There are many globulars in starburst regions (places where massive amounts of star formation took place), particularly in galaxies that are colliding and interacting. Our own galaxy has about 100 globular clusters, and their stars are very old and very metal-poor, indicating that they formed when the universe was still quite young and that few heavy elements existed in their birth clouds.

Galaxy Implications

Open star clusters are the basic building blocks of our galaxy. Their birth clouds are seeded throughout the spiral arms of our galaxy. Over the 10-billion-year lifespan of the Milky Way, clusters have formed and dissipated their stars out to the galaxy. Then, when the stars died, their recycled remains became part of the next generations of molecular clouds to form new stars. There are more than a thousand open star clusters spread through the galaxy. Among the most famous ones visible from Earth are:

- The Pleiades
- The Perseus Double Cluster
- The Beehive
- The Jewel Box
- The Southern Pleiades

Globular clusters, on the other hand, stay gravitationally bound together until they interact with something that tears them apart. When this happens, their stars are scattered through the galaxy in streams. Only the core stars remain in the original cluster, orbiting the galaxy's core.

The Seven Sisters

The Pleiades is an open star cluster that formed about 100 million years ago, making its stars fairly young by astronomy standards. It contains more than a thousand stars, and most of them are clustered into an area about forty-three light-years across. From Earth, we can see the seven brightest stars with the naked eye. Through a telescope, you can see them surrounded by a bluish cloud of gas and dust that the cluster happens to be passing through. The stars in the cluster will remain bound by gravity for other 250 million or so years as they move through the galaxy toward a region near the feet of the constellation Orion the Hunter. As they move, they will start to wander apart.

STAR BIRTH

Revealing the Secrets of Star Birth

Star birth is, as the physicist Heinz R. Pagels (1939–1988) wrote in 1985, a "veiled and secret event." Today, it's well known that star formation takes place deep inside interstellar clouds of gas and dust in stellar crèches that were once impossible for us to detect. Only after the process is complete does the light from the newborn star manage to leak out and announce to the universe that a new star has been born. It's a process that takes place in every galaxy across the cosmos, and one that has been going on since shortly after the universe was created some 13.8 billion years ago. With the advent of infrared-enabled instruments, astronomers have been able to peek into the clouds and learn more about this once-hidden process.

It Starts in the Dark

Star birth begins in a region of interstellar space filled with gas and dust called a molecular cloud. This process might ignite in a dark nebula, a cloud that is so dense that light can't pass through it. Something happens to disturb the thick, slowly moving globules of gas and dust. Perhaps a nearby supernova sends shock waves through the cloud, or another star passes nearby. The action spins the cloud and compresses it. Molecules of gas and the dust particles are crushed together, and that action causes friction heating. More and more gas and dust is pushed into this hot core, which grows more massive very quickly. As it does, its gravitational pull tugs more material in, compressing what's already in the interior. When temperatures and pressures get high enough, conditions are right for the process of nuclear

ASTRONOMY 101

fusion to begin in the core of this protostellar object. Molecules of hydrogen begin smacking together to form helium. That process releases energy in the form of heat and light, and that's what powers stars. The birth of the star is marked by the moment when nuclear fusion begins. After that, the newborn star continues to heat up; in the early phase of its life, it has gas jets streaming away from its polar regions. These help dissipate the tremendous heat built up as the star forms. If the stellar newborn has enough material remaining around it, it's possible that planets can form there.

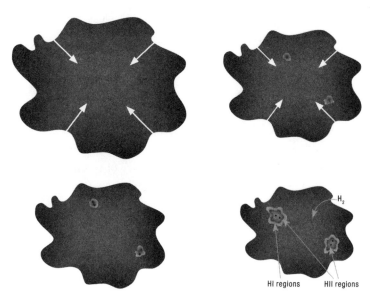

A schematic of how a dense cloud of gas and dust collapses to form a star. The core heats up as more material collects there. Wherever temperatures and pressures are high enough, a star will form. It then heats up the surrounding cloud, causing it to glow. Eventually the star eats away the star birth crèche and we see its light.

The Birth of the Sun

About 4.5 billion years ago, a small cloud of gas and dust that was part of a larger molecular cloud began to collapse in on itself. This cloud was seeded with materials from other stars as they aged and died. It's possible that some of those elements came from at least one supernova explosion, and maybe more.

Strong stellar winds from an aging star or shock waves from an explosion may have set the Sun's birthplace to spinning and coalescing. Within a few hundred thousand years, conditions at the core of the cloud were hot enough to start nuclear fusion, and a star was born. The Sun and its planets did not form alone. It was likely part of a cluster of stars that eventually scattered a few hundred million years after they were born.

Star Birth Regions in the Milky Way

Our galaxy contains many places where stars are being born. The most famous is the Orion Nebula (M42). It appears beneath the three belt stars of the constellation Orion, the Hunter, and is about 1,500 light-years away from us. At the heart of the nebula lies a collection of hot young stars. The four brightest are nicknamed "the Trapezium." These stellar newborns are heating up the surrounding clouds, causing them to glow. Such a glowing cloud is called an *emission nebula* because it emits light and heat. There are also small, disk-shaped globules of dust in many star birth regions called *protoplanetary disks*. These are places around young stars where planets can form.

The nebula surrounding the explosive binary star Eta Carinae is another well-known birth cloud in our galaxy. Nicknamed "the

Homunculus," it lies about 7,500 light-years away in the Southern Hemisphere constellation of Carina. Its clouds glow from the radiation of hot young stars. Their ultraviolet radiation is strong enough to destroy the cloud, which cuts off the supply of material to form other new stars.

Many galaxies are rippling with star birth activity. The Large Magellanic Cloud, a neighbor galaxy of the Milky Way, has a gorgeous star-forming crèche called 30 Doradus, nicknamed "the Tarantula Nebula." Even from a distance of 160,000 light-years, observers can easily spot clusters of hot young stars that have just exited their nests. They, too, are heating up the surrounding nebula and sculpting the birth cloud as they eat away at the remaining cloud of gas and dust.

The First Stars

The first stars in the universe formed from the hydrogen and helium created in the Big Bang. They were massively huge stars that began coalescing perhaps only a couple of hundred million years after the universe was born. Because their starting masses were so high, these first-borns used up their nuclear fuel very quickly, creating the first heavy elements in a process called *stellar nucleosynthesis*. When those stars reached the end of their lives, they scattered all their elements into space by means of strong stellar winds and eventually in tremendous supernova explosions. They were the first stellar objects to go through the cosmic process that continues today, spreading elements through interstellar space for the creation of new stars, planets, and in the case of our own planet, the origins of life.

Secrets of Interstellar Clouds

The interstellar clouds in which stars are born are rich in the elements needed for star formation. They're made mostly of hydrogen, with the rest being helium and a mix of heavier elements such as carbon, oxygen, and nitrogen. Where do these interstellar clouds come from? The hydrogen and helium (plus a little lithium) were created in the Big Bang 13.8 billion years ago. The other elements are cooked up inside stars or formed in supernova explosions. When stars die, much of their mass is ejected into space, where it mingles with the hydrogen and helium already there.

These nebulae can be amazing mixing pots of material. Astronomers probing these clouds have found molecules of what are called "pre-biotic" molecules that are involved in the creation and evolution of life. Our own birth cloud was rich with the compounds and molecules that are precursors to life.

STAR DEATH

How Stars Age and Die

By the standards of a human lifetime, stars seem to last forever. Even the shortest-lived ones—the massive, hot OB stars—live for a million or so years. On the other hand, dense stellar objects called *white dwarfs* spend tens of billions of years dwindling down to become cold cinders called *black dwarfs*. As they go through their lives, stars fuse elements in their cores in a process called nuclear fusion. That's what the Sun is doing right now. It's on the *main sequence*, a phase where stars spend their time fusing hydrogen in their cores. When they stop fusing hydrogen, they leave the main sequence, and that's when things get interesting. The aging and death process of a star depends very much on its starting mass. A star like the Sun dies very differently than a star that is forty times more massive.

The Ring Nebula

When the Sun dies, the material it loses to space will form a planetary nebula, similar perhaps to the famous Ring Nebula in Lyra. This object lies about 2,300 light-years away, and it has been a frequent observation target of the *Hubble Space Telescope*. The shell of material around the central star contains the elements helium, oxygen, and nitrogen. There are planetary nebulae throughout the Milky Way as well as in other galaxies.

The Death of the Sun

People often ask what will happen when the Sun dies. Let's follow the Sun's evolutionary path to the end. The Sun will continue to fuse hydrogen in its core for a few billion more years. At some point, the hydrogen will run out and the Sun will start fusing helium in its core to create carbon. That's when the Sun leaves the main sequence and becomes what's called an *asymptotic giant branch star.* It will swell up, and the carbon in its atmosphere will give it a reddish color. Eventually, the aging red giant Sun will start puffing its carbon-rich outer atmosphere into space to become part of the interstellar medium. That's a process called *mass loss.* That material will form a shell of gas and dust around the Sun called a *planetary nebula.* That term was bestowed on these objects by William Herschel because he thought they looked like planet-shaped nebulae (clouds of gas and dust) through his telescope. The hot, naked core of our Sun will be all that's left, and its intense heat and radiation will energize the planetary nebula, causing it to glow. Ultimately, the core will shrink and become a white dwarf.

Did You Know?

A nova is a nuclear explosion that takes place when a white dwarf orbits with a more massive companion, such as another main sequence star or a red giant. The companion gradually loses mass to the white dwarf. When enough material gets built up on the dwarf, it ignites and begins nuclear fusion. That causes a temporary brightening that eventually dies down, and then the process starts all over again.

Planets and Star Death

Star death is not a great experience for planets. When the Sun becomes a red giant, its atmosphere will swell up and blow away. When that happens, Mercury and Venus will probably be destroyed. What's left of Earth could be shoved out to the orbit of Mars. Mars itself will migrate beyond its current orbit and could enjoy a brief period of warmth. Eventually, the frozen worlds of the outer solar system could become warmer, possibly hosting a new wave of life. As the white dwarf Sun cools, the worlds outside its rapidly shrinking habitable zone will freeze. Eventually, tens of billions of years from now, our Sun could be a cold cinder, not even giving off light.

Massive Star Catastrophe

What about stars much more massive than the Sun? How do they die? Stars with at least eight times the mass of the Sun go through the same stages of nuclear fusion in their cores. However, they don't stop at carbon. They continue on to fuse carbon to neon, neon to oxygen, oxygen to silicon, and ultimately end up with cores of iron. At that point, the fusion reaction stops because iron takes more energy to fuse than the star is capable of supplying. When that happens, the core collapses into an extremely dense ball of neutrons. It becomes a neutron star. If the original star is very massive (say greater than twenty-five solar masses), the core can keep condensing to become a stellar black hole.

As the core is going through its changes, the rest of the star experiences mass loss when its outermost layers expand into space. Some of the star's outer layers come crashing down onto the core and then bounce back out, and that creates a shock wave. The resulting catastrophic outburst is called a *Type II supernova*. The exploding material and a shock wave rush out, colliding with material ejected earlier in the star's life. That energizes the gases in those remnants, causing them to glow. Sometimes a white dwarf star will undergo a violent explosion of its own. These are called *Type Ia supernovae* and usually result in the destruction of the star. The light from such an explosion can be used to measure distances in the universe.

The Supernova of 1054

There are remnants of supernovae scattered throughout our galaxy and in other galaxies, too. One of the most famous is the Crab Nebula, spottable through a small backyard-type telescope. It lies about 6,500 light-years away in the constellation Taurus. The light from the explosion of its massive progenitor star was first visible from Earth in the year 1054, and so its formal designation is SN1054. Chinese observers noted it as a "guest star," and it's possible that it was observed and recorded by the Anasazi culture in the American Southwest. The Crab Nebula gives off pulsations of radio signals that are coming from the neutron star spinning at its heart. This so-called *pulsar* (from the two words *pulsating star*) was discovered in 1968 when astronomers noticed the signal coming from the heart of the Crab.

BLACK HOLES

Extreme Gravity Sinks

Everybody's heard of black holes. They're popular in science-fiction literature and movies because they add great drama to the stories. In real life, black holes *are* pretty dramatic. So, what are they? Black holes are massive, dense objects with gravity so strong that nothing—not even light—can get out. That makes black holes themselves impossible to see. However, their effects *can* be detected. For example, activity around a black hole heats up its local environment, so astronomers can use instruments sensitive to infrared, x-rays, radio waves, and ultraviolet to trace what happens in the neighborhood of one of these cosmic monsters.

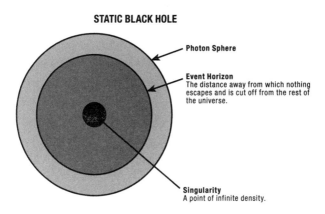

STATIC BLACK HOLE

Photon Sphere

Event Horizon
The distance away from which nothing escapes and is cut off from the rest of the universe.

Singularity
A point of infinite density.

The anatomy of a black hole. The singularity is the infinitely dense heart of the black hole. The event horizon is the "point of no return" around a black hole. The photon sphere is the region where photons are constrained by gravity to orbit endlessly.

Types of Black Holes

There are three kinds of black holes:

- *Mini black holes*, which were created in the first moments of the universe and have probably all evaporated
- *Stellar black holes*, which form during the deaths of very massive stars
- *Supermassive black holes*, which live in the hearts of galaxies

The Anatomy of a Black Hole

Stellar black holes form when an extremely massive star—say one at least twenty-five times the mass of the Sun—is blown apart in a supernova explosion. The leftover core collapses, and what's left becomes a *singularity*—a hugely dense point in space that can contain many times the mass of the Sun. The properties of a black hole, including mass, help explain its behavior. In addition to mass, black holes also have an electric charge, and most have angular momentum, or spin, which comes from the original rotation of the star that collapsed. These properties can be measured from outside the black hole, but there is no way to know about the material inside the black hole. The boundary of a black hole is called the *event horizon*. That's the region where the black hole's gravitational pull is too strong for any incoming material or light to escape. An observer outside the black hole would not be able to detect anything happening inside the event horizon.

Close to Home

The Milky Way has its very own central black hole called Sagittarius A*. It's about 26,000 light-years away from us in the direction of the constellation Sagittarius. It's a very bright radio source and also gives off x-rays. The mass of this black hole is just over 4 million solar masses. Radio astronomers have been working to measure the size of its accretion disk, and others are charting the motions of nearby stars.

We can see objects and events occurring close to the event horizon. However, the extreme gravitational pull of the mass of the black hole slows down time, and the closer those objects and events are, the slower they appear to be moving. At the event horizon, they look like they are frozen in time, taking an infinite amount of time to disappear. If *you* were swirling into a black hole, time would pass completely normally for you. But your body would take a beating. If you fell toward it feet-first, your feet would be pulled more strongly than your head, and you would be stretched into a long, noodle-like object. This is called *spaghettification*. If you could survive the trip, you'd probably never notice exactly when you passed the event horizon. You would eventually make your way into the singularity, and your atoms would be crushed along with everything else that got caught in the black hole.

A rotating black hole can have what's called an *accretion disk* around it. That's a disk of material gradually being pulled into the black hole and is one of the ways that black holes grow. As this material (gas, dust, planets, stars) whirls around in the disk, it heats up by friction. The heat is radiated away, and the disk's momentum is also funneled away through jets that stream perpendicular to the accretion disk. Activity in the disk and jets is very energetic and gives off strong emissions across the electromagnetic spectrum.

Supermassive Black Holes

The largest black holes live at the centers of galaxies and may play a role in shaping galaxies as they evolve. These supermassive black holes can have hundreds of thousands or millions of solar masses of material inside. They probably formed from black hole "seeds"—stellar-mass black holes that formed when massive supergiant stars exploded as supernovae—or from intermediate-mass black holes that were created directly from the primordial gas in the early universe. The supermassive ones continued to accrete material, growing denser as time went by. Such a black hole at the center of a galaxy would have access to a lot of matter and could grow very massive.

There are several theories about the formation of these supermassive black holes. They may come from the collapse of dense stellar clusters. It's also possible that central black holes may meld together during galaxy collisions. Even as astronomers find more and more black holes in galaxies, these strange cosmic beasts remain a fascinating and tantalizing area of study.

Black Holes, Space-Time Dragging, and Gravitational Lensing

In his Theory of Relativity, Albert Einstein (1879–1955) explained that gravity from massive objects can affect space-time. As a massive object spins, it "drags" and "bends" the local space. That warps or curves space and causes objects (and light) to fall back toward the warping mass. To think about how that works, imagine standing on a trampoline. Your mass deforms the trampoline mat downward, and a ball tossed at you would roll toward your feet. A black hole deforms space-time, and anything—including light passing by from a distant object—would be deflected slightly from its path. Depending on the viewing geometry, we might see a distorted image of the background galaxy. That process is called black hole gravitational lensing.

GALAXIES

Stellar Cities

Galaxies are huge collections of stars, gas, dust, and dark matter all bound together by gravity. Like snowflakes, no two galaxies look exactly alike. They range from elliptically shaped globes and gorgeous spirals to irregular collections of stars, gas, and dust. These shapes— what astrophysicists call *galaxy morphologies*—provide clues to the lifestyles and evolutionary history of these massive stellar cities.

Taking Galactic Inventory

All galaxies have stars. Most of those stars are in multiple-star systems such as binaries, triples, and clusters. In spiral galaxies, many of the stars are arrayed in bright arms that extend out from the central galactic core. These arms are the sites of much of the star formation that goes on, and they are interspersed with clouds of gas and dust. In the Milky Way (which is a barred spiral), astronomers are finding many stars with planetary systems, and so it's reasonable to assume that plenty of stars in other galaxies also have planets.

When a galaxy's stars die, they leave behind clouds of gas and dust that spread out to interstellar space in the galaxy. They also create neutron stars, white dwarfs, and pulsars—the exotic end-points of stars. The clouds of gas and dust mingle with other gases in the interstellar medium, and eventually these nebulae become the birthplaces of new generations of stars.

Many galaxies have supermassive black holes at their hearts. The accretion of matter into massive disks swirling around the black holes in galactic cores sometimes creates jets of material that stream into intergalactic space. If a galaxy has activity in its core that sends out emissions across the electromagnetic spectrum, that region is called an *active galactic nucleus*. Not all of these objects have jets, but they all show activity in the central regions that emit light across the electromagnetic spectrum.

Finally, in recent years, researchers have found that galaxies seem to have hidden mass in the form of a substance called *dark matter*. What this material is has been a mystery, but it does exist, and its gravitational effect on galaxies is detectable and measurable.

Observing Galaxies

There are two ways to learn about galaxies: first, we can observe our own Milky Way (from the inside, of course), and second, we can study distant ones. There are only three galaxies that can be seen with the naked eye from Earth (besides our own). These are the Andromeda Galaxy and the two satellite galaxies of the Milky Way called the Large and Small Magellanic Clouds. These have been observed since antiquity, although it took until the twentieth century for astronomers to determine exactly what they were. All other galaxies can only be seen using telescopes. The most distant galaxies require extremely high magnification and specialized instruments to detect.

Galaxies are part of a group of celestial objects often referred to as "deep sky" objects. "Deep" simply means "more distant," but it can also refer to objects that are intrinsically dim and hard to observe. It's necessary to aim telescopes and imaging systems at such objects for long periods of time in order to gather the greatest number of photons. Such images are called *deep fields* or *deep sky surveys*. The longer astronomers look, the more of these objects they find. Many galaxy studies are done in such surveys; recent ones performed using *Hubble Space Telescope* have found a backdrop of very distant early galaxies.

Galaxy studies are a recent development in the history of astronomy research. In the eighteenth century, observers such as Charles

Messier (1730–1817) and later on William Herschel (among others) began finding dim, oddly shaped nebulae through their telescopes. It wasn't until the early twentieth century that people began seriously studying the light from these "spiral nebulae" through spectroscopes (instruments that split light into a rainbow of colors called the *electromagnetic spectrum*). The data implied that these objects were moving very quickly and could be well outside the Milky Way. That was puzzling since, until that time, it was assumed that everything in the sky was somehow gravitationally bound to our Milky Way.

That all changed in the 1920s when astronomer Edwin Hubble (1889–1953) looked at the Andromeda Galaxy through the 100-inch Mount Wilson telescope. He measured the light coming from stars called Cepheid variables, and his calculations confirmed that Andromeda lay well outside the Milky Way. They showed the universe was much larger and more complex than expected. In 1926, Edwin Hubble came up with a scheme to classify galaxies by their shapes, which is the basis of how galaxies are sorted today. His work is often referred to informally as the "Hubble Tuning Fork."

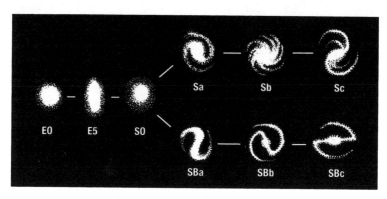

The famous Hubble "tuning fork" diagram shows the basic classification of galaxies by their shapes (morphologies).

Shapes of Galaxies

Galaxies come in several basic shapes, and astronomers use those shapes to classify them. At least two-thirds of known large galaxies are *spirals* like the Milky Way, making them the most common type. *Elliptical galaxies* are the next most common. These are globularly shaped and don't seem to have any spiral arms. Finally there are the *irregular galaxies*—amorphous blobs of stars, gas, and dust like the Large and Small Magellanic Clouds. Most smaller galaxies in the universe are dwarf ellipticals and dwarf irregulars.

Galaxy Networks

Galaxies exist in collections, ranging from groups and clusters to superclusters. Actually, most galaxies are gravitationally bound together with others. For example, the Milky Way belongs to a collection of about fifty-four galaxies called the Local Group. It stretches across roughly ten million light-years of space and includes Andromeda and the Magellanic Clouds. The Local Group itself is part of a larger collection of about 100 galaxy groups and smaller clusters called the Virgo Supercluster. This local supercluster covers about 110 million light-years and is part of a network of many superclusters. All the galaxies and their clusters and superclusters link together to form the *large-scale structure of the universe*.

How Many Galaxies Are There?

Galaxies fill the universe for as far as we can observe, arranged in groups, clusters, and superclusters. There are estimated to be at least 180 to 200 billion galaxies out there—and perhaps as many as a trillion!

GALAXY FORMATION

From the Big Bang to Today

Galaxies—where do they come from? How long do they exist? These are questions that astronomers are only just now beginning to answer, using observations of the most distant "baby" galaxies that were born right after the Big Bang. They are also studying our own Milky Way and its continuing evolution. It turns out that galaxies didn't start out as the gorgeous and fascinating objects we see today. They were built over billions of years by systematic mergers and collisions.

Galaxies are some of the oldest structures in the universe. Most of them began like our own Milky Way—getting their start as cosmic seeds in the earliest epochs of the universe, not long after the universe began in an event called the Big Bang. This occurred 13.8 billion years ago and was the simultaneous creation of our universe and the beginning of an expansion of space and time that continues to this day. For the first few hundred million years, things were hot and opaque and the universe consisted of a soup of primordial atomic particles. Eventually, as this nascent cosmos expanded, it cooled. Tiny fluctuations in the density of this basic soup became the seeds of galaxies, helped along by the gravitational influence of dark matter.

About 400 million years after the Big Bang, the first stars began to shine in those infant galaxies, which were little more than shreds of starlit matter. Eventually these primitive galactic shreds began to combine. They collided with each other to form larger collections of stars. As they did, successive generations of stars were born, lived, and died. It was this way with the infant Milky Way, which first began forming at least 11 to 12 billion years ago. Its earliest stars eventually died and seeded the galactic environment with materials that got recycled into new generations of stars. Some of the remnants of our galaxy's earliest stellar population still exist as slowly cooling white dwarfs in the galaxy's gassy halo. However, there *are* other stars that are as old as our galaxy—they most likely formed in globular clusters that we see hovering and orbiting in the galaxy's halo.

Today the Milky Way is still cannibalizing neighbors that stray too close. There's a small spheroid-shaped elliptical galaxy called the Sagittarius Dwarf that is slowly spiraling into our galaxy as it orbits. As it goes past, it leaves behind streams of stars that are metal-poor. This indicates they formed early in the history of the universe, and the exact metal content identifies them as being part of the Sagittarius Dwarf. The discovery of this ancient galaxy and its stars in close proximity to the Milky Way was one of the key pieces of evidence that astronomers needed to show how collisions shaped the Milky Way.

Disks of Light

The Milky Way is a disk galaxy. It spins rapidly, and most of its stars, gas, and dust have collapsed into a flat disk. In a galaxy collision, that disk can be shredded pretty easily. That's why some spiral galaxies seem to have warped disks. Close encounters with other galaxies cause gravitational disturbances that pull the spiral disks out of shape.

The inexorable pull of gravity might also bring the Large and Small Magellanic Clouds into our galaxy in a few billion years. It appears that these two neighboring galaxies have already been interacting with each other. There's a river of high-speed gas connecting them called the Magellanic Stream, and it may be left over from a close encounter between the pair more than 2 billion years ago that triggered huge bursts of star formation. Such starbursts are important outcomes of galaxy collisions. For example, the Milky Way and Andromeda (our nearest spiral neighbor) grew through ever-more-complex galaxy mergers and near-misses that took millions and millions of years to complete. When the participating galaxies hit head on, in a full merger, their stars mingled but didn't collide. As with the recent encounter of the Magellanic Clouds, the close encounters also sent shock waves through the clouds of gas and dust, and that spurred huge bursts of star birth activity. So, as galaxies evolve and merge, their populations of stars are enriched by ongoing bouts of star formation.

The evolution of galaxies is a constant work in progress. For example, the Milky Way and Andromeda are bound together by a common pull of gravity. They're approaching each other at 110 kilometers (68 miles) per second. In about 5 billion years, they will actually pass through each other. They'll mingle stars, but more importantly, they'll draw some of each other's gas and dust out into long, intergalactic, star-forming streamers. Over several billion years, the two will perform a delicate cosmic dance—passing through each other several times before ending up as a giant elliptical galaxy.

Elliptical Galaxies

Ellipticals are galaxies in the most advanced state of evolution. They are the largest, most massive collections of stars that are moving in random orbits. These galaxies form directly from violent collisions of smaller galaxies, and there's no chance for spiral arms to exist in them. What's more, ellipticals all have black holes at their hearts, and some of them are shooting jets of material out into intergalactic space. The elliptical galaxy M87, for example, has a high-speed jet of superheated matter streaming away from the region around its central 6-billion solar-mass black hole.

How Galaxies Interact

Galaxies can interact in various ways as they travel through space:

1. **GALAXY COLLISION:** when two or more galaxies directly collide one or more times
2. **CANNIBALIZATION:** when one galaxy merges with another and creates an elliptical or irregular galaxy
3. **INTERACTIONS:** these usually take place between a large galaxy and its smaller satellites (which could be dwarf galaxies)

THE MILKY WAY

Our Home Galaxy

If you've ever been outside at a good dark observing place, you might have seen the Milky Way. It almost looks like a cloud, and in a way it is. It's a star cloud, and it's also how our galaxy looks from the inside. If you could somehow travel outside our galaxy and look back at the Milky Way, it would resemble a giant pinwheel of light with spiral arms winding around a bright core that has a bar of light extending across the center. Our galaxy has been around for about 11 billion years, forming from smaller clumps of stars into the gigantic stellar city we inhabit today.

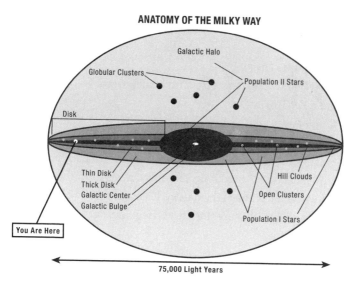

ANATOMY OF THE MILKY WAY

Galactic Halo

Globular Clusters

Population II Stars

Disk

Thin Disk
Thick Disk
Galactic Center
Galactic Bulge

Hill Clouds

Open Clusters

Population I Stars

You Are Here

75,000 Light Years

The anatomy of the Milky Way Galaxy. We live in the plane of the galaxy, about 26,000 light-years from the core.

Milky Way Facts

1. There are more than 400 billion stars in our home galaxy.
2. The Milky Way measures about 120,000 light-years across.
3. Our galaxy rotates once every 220 million years.

A Galactic Tour

The central region of the Milky Way is densely populated with stars and a black hole called Sagittarius A*. You can't see the core of our galaxy because it's hidden behind clouds of gas and dust, but it's a very busy place. The innermost part of the core has some of the oldest stars in the galaxy packed into a roughly spherical area about 10,000 light-years across. Extending out from the core is a bar of gas and stars that connects to the two major spiral arms of the galaxy called the *Scutum-Centaurus Arm* and the *Perseus Arm*. Other, smaller arms extend out from the center:

- The Sagittarius Arm
- The Near 3KPC Arm (short for "three kiloparsecs")
- The Far 3 KPC Arm
- The Outer Arm

The Sun is located about 26,000 light-years out from the center, in a "sub arm" called the Orion Spur. The spiral arms form the galactic disk, which is the flat plane of the galaxy. It's made mostly of stars and clouds of gas and dust that are the sites of ongoing star formation.

Part of a Galactic Company

The Milky Way travels through the universe as part of a group of fifty-four galaxies called the Local Group. Other members include the Large and Small Magellanic Clouds and the Andromeda Galaxy.

Surrounding the disk is the galactic halo, which encloses a region of space about 200,000 light-years across. It contains very old stars and many globular clusters, which are gravitationally bound collections of old stars. The globulars orbit the core of the galaxy and are thought to have formed about when the galaxy did. Beyond the visible halo is a spherical concentration of unknown material called dark matter. Recent studies of our galaxy using x-ray data show that the galaxy travels the universe embedded in a bubble of very hot gases.

Forming the Milky Way Galaxy

The formation history of galaxies is a story that astronomers are really just starting to understand. Here's how the story goes: A few hundred thousand years after the Big Bang, the universe was a distributed mass of matter that was denser in some areas than in others. The first stars began to form inside these "overdensities," and these stars became the seeds of our own galaxy and of the globular clusters that swarm around its core. Billions of years later, the spherical-shaped early Milky Way had a critical mass of stars and other material, and it began to spin. This motion caused it to collapse into the disk shape we know today.

Humans and the Milky Way

Cultures throughout history have referred to the part of the Milky Way we see from Earth by some very poetic names. In Latin it was called *Via Lactea* (Milky Way), which is similar to its name in Spanish: *Via Láctea*. The Korean name for this band of light is "Silver River." In South America, ancient Incan sky watchers saw their portion of the Milky Way as a Celestial River, flowing past constellations of llamas, condors, and other creatures familiar to them. In Australia, aboriginal people believed it was a river to the underworld. Today the Milky Way is known as a gorgeous patch of sky to observe, providing myriad stars and a wealth of star clusters, nebulae, and other objects for people to study.

The Milky Way continued to grow and change through galaxy mergers. Over time, these collisions helped develop the spiral arms. In fact, our galaxy is still assimilating stars from a merger with small dwarf spheroidals and is pulling material from the Large and Small Magellanic Clouds, two satellite galaxies. The process isn't through yet. The Milky Way is moving through space at an estimated velocity of 630 kilometers per second, and it's headed for a merger with the nearby Andromeda Galaxy. That will take place in about 4 to 5 billion years, and the two galaxies will ultimately mingle stars, gas, dust, and possibly their central black holes. The process will radically change the shape of the resulting merged object, triggering great bursts of star formation and creating the so-called Milkdromeda Galaxy, a giant elliptical that will ultimately combine with the other galaxies in the Local Group.

The Orion Nebula is one of the closest and best examples of a star birth region in our galaxy. It lies 1,500 light-years away and contains more than three thousand stars of various ages and sizes, plus many regions where stars are still being formed.

Optical

Composite

Infrared & X-rays

Infrared

X-rays

The Andromeda Galaxy is the closest spiral galaxy to the Milky Way. The views here from eight different observatories in various wavelengths of light show details of its structure, star-forming regions, and the distribution of stars, gas, and dust throughout the galaxy's spiral arms and central bulge.

Photo Credits: Spitzer Space Telescope/Digital Sky Survey/Jason Ware/GALEX/Effelsberg Radio Telescope/ROSAT/IRAS/ISO

The Pleiades open star cluster is well known to both amateur and professional astronomers, and easily spotted in the sky from November to April. It lies about 440 light-years away and contains more than a thousand stars. It is currently passing through a cloud of gas and dust, which gives the stars a wispy appearance.

Globular clusters contain some of the oldest stars in our galaxy and formed as the galaxy was assembling itself. This is a view of the central region of M13, a globular that lies in the constellation Hercules and contains more than 100,000 stars. The cluster is very densely packed with aging giant stars.

This *Hubble Space Telescope* image shows a very distant galaxy cluster called MACS 1206. Its light is being distorted by the gravitational influence of dark matter within the cluster, causing the galaxies to appear warped, as if glimpsed through a distorting lens. Astronomers surveyed a number of such galaxy clusters to understand the distribution of dark matter in the universe.

Photo Credit: NASA/ESA/M. Postman/STScI

This group of interacting galaxies is called Arp 273. Galaxy collisions and interactions are important influences on a galaxy's evolution. In this view, the large spiral galaxy at the top is being distorted by the gravitational influence of the galaxy below it as they pass by one another. This close passage spurs intense star formation activity, which results in batches of hot young stars, indicated by the blue regions.

The Crab Nebula is one of the most famous supernova remnants in the galaxy. It first appeared in our night sky in the year 1054, heralding the death of a supermassive star. It hurled much of its mass out to space, leaving behind a rapidly spinning neutron star called a pulsar. *Hubble Space Telescope* mapped the distribution of different gases in the cloud.

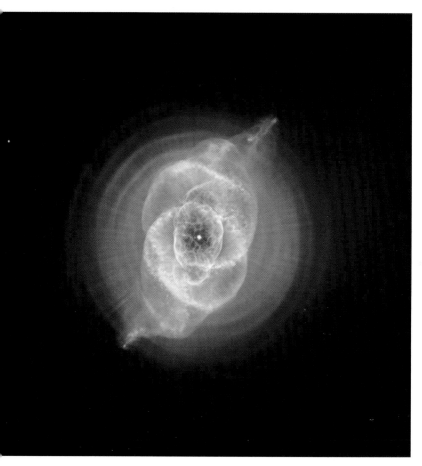

How will the Sun die? This *Hubble Space Telescope* image of the Cat's Eye Nebula shows one possible scenario. It will expel much of itself into space, creating a cloud surrounding a slowly cooling white dwarf star that will light up the cloud.

This is one of *Hubble Space Telescope*'s deepest images of distant galaxies, taken in near-infrared light and showing some of the earliest galaxy objects in the universe. They appear as they did when the universe was only about 600 million years old.

Why a Bar?

The heart of our galaxy contains a bar of stars, along with some gas and dust, and there is a great deal of interest in how it formed and what its function is. It was discovered when astronomers studied light from stars moving near the center of our galaxy. They found that one end of the bar is pointed almost directly at the solar system. The bar itself rotates like a roller while the rest of the galaxy is turning more like a wheel. There are hundreds of stars with long, looping orbits around the center of the galaxy, and the bar seems to act like a giant mixer, churning up the motions of nearby stars in the core and possibly affecting those out in the disk.

ACTIVE GALAXIES AND QUASARS

Galactic Monsters

There are galaxies in the distant universe figuratively shouting their presence out through the megaparsecs through massive blasts of emissions across the electromagnetic spectrum. These are some of the most luminous objects in astronomy. What could possibly be happening in their hearts that would let us detect their radiation from Earth? For an answer, let's journey back to a time when these massive galaxies were very young. Stars in their cores were packed tightly together, and gravitational interactions caused violent stellar collisions. In each galaxy, a black hole formed at the core, and it swallowed up stars and possibly other black holes created when massive stars died. In time, this supermassive black hole had the mass of perhaps a billion suns! If the galaxy collided with a neighboring one, which happened frequently in the early universe, there was even more material and perhaps another black hole, all available to feed the hungry monster.

In time, the black hole at a galaxy's core contains ample material around it in the form of hot gas, stars, and other material filling the interstellar medium. The matter spirals toward the black hole, creating a flat, pancake-like structure called an accretion disk. Some disks at the hearts of galaxies are threaded through with strong magnetic fields that get twisted as the disk spins around. Material caught in the disk gets hotter and hotter through magnetic heating and the friction of particle collisions. The heat escapes as energy, and it gives off strong radio and x-ray emissions. Some of the material can also escape as massive jets traveling perpendicular to the disk and out into space. Some of these jets blast out at close to the speed of light! Today astronomers classify such busy regions in the hearts of galaxies as *active galactic nuclei* or *AGN*. They are very compact and very bright in almost all wavelengths of light. Nearly all can be explained by the interaction of a supermassive black hole with material in nearby space.

Cosmic Evolution and AGN

Astronomers know that greater numbers of AGN existed when the universe was much younger and more conducive to forming such objects. Perhaps there were more black holes to merge at the centers of galaxies. It's also possible the gases needed to create stars were more plentiful back then, giving rise to the crowded conditions at galactic cores that would foster the births of black holes. As well, there were likely more galactic interactions since the early universe contained many smaller galaxies.

Many AGN lie at very large distances from us, at *high redshift*. In particular, the existence of very distant Seyfert galaxies giving off gamma-ray glows indicate such objects exist everywhere in the universe.

Quasars!

Quasars (short for "quasi-stellar radio sources") are the most energetic and distant active galactic nuclei known. As their name implies, these objects were discovered through their strong radio emissions, but they're bright in visible light, too. Some give off strong x-ray emissions as well, which gives an idea of the high temperatures in the central cores of these galaxies.

Like other AGN, a quasar is powered by a supermassive black hole surrounded by a monstrous accretion disk funneling material in at a prodigious rate. The brightest known quasar must gobble up the equivalent of a thousand suns each year to keep the lights on. Eventually, an AGN's core black hole runs out of the fuel it needs to sustain the bright lights and emissions. When this happens, the central region quiets down. Not every galaxy with a central black hole is a noisy monster. Our own Milky Way has a relatively quiet black hole compared to other, really active galaxies.

Discovering Active Galactic Nuclei and Quasars

These active objects were discovered early in the twentieth century when people began using spectrographs to study the hearts of galaxies. They found bright emission lines, which indicated the areas were active and superheated. Astronomer Carl Seyfert (1911–1960) first wrote about these so-called "active galaxies" in 1943. Their strong emissions indicated something very energetic was going on the central cores. Eventually they became known as *Seyfert galaxies*.

In the 1950s, radio telescopes began detecting strong radio sources in the same regions. Often enough, there would be a galaxy with strong radio sources on either side of it, and eventually astronomers figured out these were jets emanating from the cores of sometimes normal-looking galaxies.

Really, Really Bright!

The brightest quasars produce the same amount of light as 26 quadrillion Suns!

Quasars took a bit longer to explain. They appeared bright, almost starlike, and their brightnesses would sometimes vary over a few days. However, it was hard to tell what they were associated with, and they were almost always very, very far away. Eventually, Dutch astronomer Maarten Schmidt (1929–) looked at emission lines from a quasar called 3C 273. He pointed out these lines were coming from energized hydrogen atoms and were highly redshifted (that is, the spectra were shifted very far toward the red end of the spectrum). 3C 273 was not only very bright, but also very far away. Eventually people realized that quasars are powered by extremely massive black holes consuming huge amounts of material.

Today scientists are exploring the correlation between a galaxy and the mass of its central black hole. In particular, they are studying how the black hole's role in gobbling up material eventually robs a galaxy of its chance to make batches of new stars. The central engine's prodigious appetite seems to be causing its own eventual downfall as it runs out of fuel to keep accreting mass.

Types of Active Galaxies

Active galaxies are characterized by the emissions they give off and whether or not they emit jets from their cores. Here are a few of the most common types.

1. *Radio-quiet*: very dim, quiet galaxy cores with radio quiet (for now) black holes; they may be bright and active in other wavelengths of light
2. *Seyfert galaxies*: medium-mass black holes accreting material and giving off x-rays and gamma rays
3. *Quasars*: high-mass black holes accreting material; some emit radio emissions while others emit only optical light

4. *Blazars*: high-mass black holes with a jet pointing toward Earth
5. *Radio galaxies*: high-mass black holes with large areas that give off strong radio emissions and have massive jets streaming superheated material into space

Microquasars

Quasars generally exist in far distant reaches of the universe, but in recent years, radio-loud jets have been found emanating from double-star systems in our own galaxy. These powerful jets appear to be moving faster than the speed of light—a property called "superluminal motion." It turns out these jets are aimed directly at us, which contributes to the illusion of faster-than-light travel. The power output of these objects is incredibly high, and astronomers now call them microquasars because they behave almost exactly like their larger counterparts in the centers of distant galaxies.

DARK MATTER

The Mysterious Stuff of the Cosmos

There's an invisible substance out there in the universe that is unde-tectable except through its gravitational influence on the matter we can see. It's called *dark matter*. To understand how something unseen can be so influential, let's first talk about the matter we *do* know about.

Every object in the universe is made of little bits of matter called atoms. These atoms comprise the basic building blocks of the chemical elements such as hydrogen, helium, carbon, nitrogen, oxygen, silicon, and so on. They combine, along with energy, to make everything we observe.

Keeping the Universe Together

What holds the atoms together? The "force" does. In the atomic and subatomic realm, atoms are governed by two interactions called the *strong nuclear force* and *weak nuclear force*.

The strong nuclear force works on the pieces that make up the atom's nucleus: the protons and neutrons, which are themselves made of smaller particles called *quarks*. The strong nuclear force keeps the quarks from pushing apart and destroying the nucleus.

The weak nuclear force changes one type of quark into another, which causes radioactive elements (such as uranium) to decay. Radioactive decay is a power source inside planets and moons, and the elements involved are created when stars die.

The weak nuclear force also enables a process called nuclear fusion, which occurs in the center of the Sun and other stars. There, atoms of one element fuse together to become a second, heavier element. In the process, heat and light are emitted.

The *force of gravity* acts over longer distances and on larger, more massive objects. It's universal—it acts the same way everywhere. Every object has mass—from the smallest specks of cosmic dust to the most extensive galaxy cluster. The mass of one object attracts other objects, and the strength of that attraction depends on the distance between them. Gravitational interactions bind together the masses of Earth and the Moon, the Sun and its planets, the stars in the Milky Way, and galaxies together in clusters and superclusters.

All of the objects just mentioned are lumped together in a category called *baryonic matter*—material made of protons and neutrons. It reflects or emits radiation. That includes you, me, stars, planets, galaxies, and clouds of hot gas inside and around galaxies. The strange thing about baryonic matter is, even though it makes up everything we can see, when you count up all the matter that exists in the universe, it only makes up 16 percent of *all* matter. The rest—about 84 percent—is dark matter.

What is dark matter? Nobody knows for sure yet. It's not even clear whether it acts the same as baryonic matter.

Possible Dark Matter Candidates

1. *Cold dark matter*: an unknown substance that moves very slowly
2. *WIMPS*: weakly interacting massive particles (could explain cold dark matter)
3. *Baryonic dark matter*: black holes, neutron stars, brown dwarfs
4. *Hot dark matter*: highly energetic matter moving at close to the speed of light

So if dark matter is dark, how did astronomers ever come to know of its existence? If you take all of the stars, gas, and dust in a galaxy, for example, and use what you *see* to calculate its mass, the galaxy turns out to be far more massive than it should be if you were only counting those objects. Something doesn't add up. That is, there's something *there* that you can't see, and it has mass.

Astronomers noticed this discrepancy with many galaxies and clusters, and they also discovered differences in the velocities of stars as they orbited around inside a galaxy. There should be enough gravitational influence from the mass in a galaxy to keep its stars orbiting at well-predicted velocities and to keep the outermost ones from flying off into space. However, the predicted velocities of stars on the outskirts of a galaxy are not what astronomers expected them to be. They should be moving at a much different rate than the stars closer to the galaxy's center. This implies there's more mass (with its resulting gravitational pull) that isn't being detected and that mass is affecting the motions of the stars.

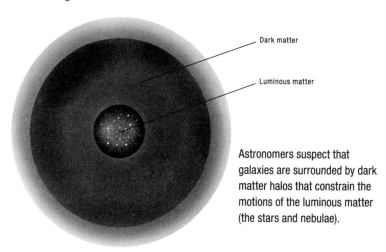

Dark matter

Luminous matter

Astronomers suspect that galaxies are surrounded by dark matter halos that constrain the motions of the luminous matter (the stars and nebulae).

Where Is the Dark Matter?

In early 2012, physicists looking into the problem of dark matter used galaxy distribution data and gravitational measurements to make computer models of the dark matter universe. Their models showed that this invisible substance extends out from galaxies far into space, and perhaps even rubs up against the dark matter components of neighboring galaxies. This means that each galaxy is really a collection of both luminous (baryonic) matter and dark matter. Furthermore, it means that dark matter itself forms a web that permeates the entire universe.

Dark Matter and the Universe

Dark matter can affect galaxies and galaxy clusters throughout the universe. Its existence can tell astronomers if the universe will keep expanding, stay static, or perhaps even shrink. Cosmologists (people who study the origin and evolution of the cosmos) used to think in terms of an open universe, a closed universe, and a flat one. An open universe is one that keeps expanding, a closed universe may contract or shrink, and a flat universe stays static. Mass plays an important role in determining what kind of universe we have, so it's important to know how much matter there is.

Today, the picture is more complicated—the universe is considered flat in terms of matter, but it's also expanding because of something called *dark energy*, which is the most dominant property when figuring out the fate of the cosmos. When you add dark energy into the mix, it turns out that the universe is about 75 percent dark energy, 24 percent dark matter, and just under 5 percent baryonic matter. The dark energy is speeding up the expansion rate of the universe, so in this scenario, the universe is likely to keep expanding indefinitely. However, recent studies of the newly discovered Higgs Boson, a sub-atomic particle, suggest that our universe *could* have a finite lifespan that will come to an end many tens of billions of years from now.

What Is Dark Energy?

People often think of space as empty. This is one of the great misunderstandings of astronomy. In fact, there's a lot of stuff in space. Atoms of gas and small particles of dust float in the so-called "void." And now, it turns out there is something called dark energy. It's not so much "stuff" as it may be a property of space—think of it as space having its own supply of energy. As space expands, more of its energy appears. That energy is what's driving the expansion of the universe. This may explain why astronomers have found that the universe is expanding faster than it should be (based on a calculation of all the mass it contains and the gravitational pull of all the material in it).

THE STRUCTURE OF THE UNIVERSE

Cosmic Objects and Their Distances

The universe is populated with objects that are in motion. Take Earth, for example. It orbits the Sun, but it is also moving through space as part of the solar system. The solar system is in motion around the center of the Milky Way Galaxy. The galaxy itself is moving through space along with other galaxies as part of a collection called the Local Group. In some places, galaxies orbit each other within clusters and associations, but all of those galaxies are also moving through space as part of the expansion of the universe that began some 13.8 billion years ago. The universe is also huge. It began as a small point and is now measured at about 93 billion light-years across!

All the matter the universe contains (both baryonic and dark matter) is distributed in a lacy network that stretches across those billions of light-years of space. In between the strands of the network are cosmic voids filled with dark matter. Finally, everything is affected by dark energy as it speeds up the expansion of the universe.

Climbing the Cosmic Distance Ladder

Distances are important in the cosmos. Obviously people explore the closest objects such as the planets and nearby stars and nebulae more easily than the distant stars and galaxies. The farther out we look, the further back in time we see, and so distance becomes an exercise in studying the history and evolution of the cosmos and the objects it contains.

Earth and the Sun are 1 astronomical unit (AU) apart. That's 150 million kilometers. At the AU level, our exploration is limited to the solar system, which is pretty small in the cosmic order. More distant objects are referred to as lying light-years away. A light-year is the distance light travels in a year at 300,000 kilometers per second. The nearest stars are about 4.2 light-years from the Sun. One of the nearest starbirth regions is 1,500 light-years away. Astronomers also use the term *parsec*, which is equivalent to 3.26 light-years. Multiples of thousands of parsecs are called kiloparsecs. Measurements in these units get us out to the objects in the Milky Way. So, for example, the center of our galaxy is about 26,000 light-years away, which can also be expressed as 8,000 parsecs, or 8 kiloparsecs.

Standard Candles

How do we know how far away something is? There's a very useful way to measure the distance to something quite distant in the cosmos. Objects in the universe that emit light at a known and predictable brightness are used as standard candles. To calculate their distance, observers use the inverse-square law, which says that the luminosity of an object is inversely proportional to the square of its distance. (Luminosity is a measure of the total amount of energy emitted by an astronomical object.) This inverse-square law says that something looks dimmer the farther away you are. If you know the object's luminosity you can calculate the distance.

The most common standard candles in use are Cepheid variable stars (which pulsate in a steady, predictable rhythm over time), certain types of stellar explosions called Type 1a supernovae, and quasars (bright distant outbursts in space). Red giants and planetary nebulae can also be standard candles. These standard candles exist in the Milky Way and elsewhere in the universe, which allows us to measure the distances to those stars—and thus to their galaxies.

Beyond Our Galaxy

Outside the Milky Way, astronomers deal in huge distances—measured in millions or billions of parsecs, called megaparsecs and gigaparsecs, respectively. These are often referred to as cosmological distances to bodies such as galaxies. Astronomers can measure these distances using the standard candles mentioned above. Another method, called the Tully-Fisher relation, compares the intrinsic luminosity (that is, how bright it is) of a spiral galaxy to the orbital motion of stars around the center of the galaxy. To get this "rotation velocity," astronomers study their light through a spectroscope—an instrument that splits light into its component wavelengths. The spectra show how fast the stars are moving as the galaxy rotates.

Sometimes changes in surface brightness (that is, the brightness across an extended object like a galaxy or a nebula) can also be used to determine its distance. This surface brightness method works well with objects more than 100 megaparsecs away.

How the Universe Got to Be That Way

Astronomers detect and study a universe filled with light-emitting objects. How did everything get to be the way it is? A few hundred thousand years after the Big Bang, the infant universe was filled with matter distributed across expanding space. At that time, the universe was much smaller than it is now because it hadn't expanded very far. In some places, the density of the matter was a tiny, tiny percent higher than in neighboring regions. In those areas, the expansion of the universe was a little slower, allowing the areas of higher density to grow and become more dense. The first stars were formed in those areas of "over-density," and the regions where they lived and died were the seedbeds of the first galaxies.

From those tiny density variations in the early universe, the largest-scale structures in the universe—the galaxies and galaxy clusters—grew and clumped together under the force of their mutual gravitational attractions. These clusters and superclusters are arrayed in walls, sheets, and filaments of matter that seem to be draped around vast, empty-looking voids. Today people are interested in the role that both dark matter and dark energy played in the creation and ongoing evolution of the stars, galaxies, galaxy clusters, and superclusters that make up this cosmic web of existence.

The fact that so many superclusters exist proves that matter in the universe is not evenly distributed and that it hasn't been since these clusters first began forming. Today, these structures stretch for hundreds of millions or even billions of light-years, and yet they define only a small percentage of the universe that we can detect.

Galaxy Clusters and Superclusters

Galaxies do not roam the universe alone. All of them, including our Milky Way, are part of clusters—large associations of galaxies held together by gravity. Some of these clusters may have a few dozen to a few thousand galaxies, while others can have tens of thousands. Groups and clusters of galaxies are all part of larger collections called superclusters, which can have tens of thousands of members. The best example is the Perseus-Pisces Supercluster. It lies more than 250 million light-years away and is made up of several groups and clusters of galaxies.

From the Small to the Large

Matter in the universe is clustered in a hierarchy. Here's a list of cosmic objects, ranging from small to large.

- Planets
- Stars
- Galaxies
- Groups of galaxies
- Galaxy clusters
- Superclusters (clusters of galaxy clusters)
- Voids between clusters and superclusters
- Filaments of galaxies outlining the voids

GRAVITATIONAL LENSES
Nature's Long-Distance Telescopes

In 1979, astronomers using a telescope at the Kitt Peak National Observatory found something strange in one of their images. It looked like a pair of identical quasars sitting side by side and fairly close together. They were quickly dubbed the Twin QSOs. Quasars are bright, very distant point-source objects that put out prodigious amounts of light and are now known to be extremely active cores of distant galaxies. It seemed weird that there would be two so close together. So the astronomers looked at the Twin QSOs with radio telescopes to see if there really were two quasars out there.

It turns out the double quasars are really two images of one distant object that lies behind a massive galaxy cluster. The combined gravitational pull of all the galaxies in that cluster is enough to deflect light from the quasar as it passes by, and that gravitational lensing is what creates two quasar images.

Gravity's Secret Role

Physical bodies all have mass. Each mass exerts an attractive force on other masses. The more massive the body, the stronger its gravitational force. Gravity and velocity keep planets in orbit around the Sun, moons going around planets, and galaxies orbiting other galaxies. So, how do they work?

We all owe our understanding of orbits and gravity to Sir Isaac Newton, an English mathematician in the seventeenth century who came up with a universal law of gravitation. Gravity, he said, is a force that acts on all objects in the universe. You can calculate the force of gravity between two objects if you know their mass and their

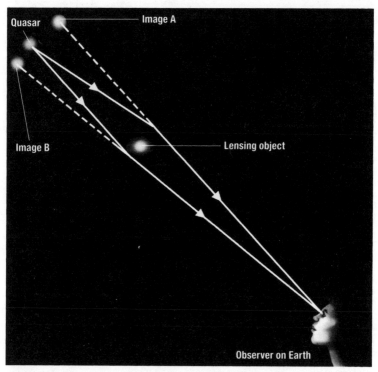

Quasar

Image A

Image B

Lensing object

Observer on Earth

This is the geometry of a gravitational lens. Light travels from the distant object, but its path is distorted by the gravitational pull of a massive object along the way. An observer sees more than one image of the more distant object.

distance from each other. The closer the objects are together, the stronger the force of gravity that attracts them to each other. The farther apart they are, the weaker the gravitational pull.

There are very detailed textbooks on gravity and its role in the universe. However, what's most important to know about gravity in astronomy is that it influences everything from star formation and galaxy evolution to the orbital mechanics of objects in our solar system to the path that light takes through the universe.

Applying a Gravitational Lens

Any distribution of matter in space can act as a lens. The bigger the mass, the more gravitational distortion it creates. Astronomers knew from Einstein's work that this was true, and some even predicted that this effect would be produced by galaxy clusters. For a gravitational lens to work, you need several things:

- A source—such as a quasar or a distant background galaxy
- Lensing material—anything from a star to a distant galaxy cluster
- An observer
- The images the observer detects

It's All Relative

Space and time are interesting things. They can both be affected by matter—particularly large amounts of matter that have strong gravitational influences. This is the basis for the work that Albert Einstein did, spurred on by a solar eclipse that occurred in 1919. He predicted that light rays from distant stars would be bent as they passed by the Sun due to the Sun's gravitational influence. The eclipse blocked sunlight, allowing observers to see stars they normally wouldn't see, and they succeeded in measuring a tiny shift in light due to gravitational lensing. This observation led Einstein to publish work describing how the mass of an object curves local space-time, thus forcing light rays to bend ever so slightly. The 1919 eclipse produced the first experimental confirmation of gravitational lensing. Today, there are many gravitationally lensed objects that have been observed, ranging from stars to distant quasars.

The Einstein Cross

Out in the distant universe—more than 8 to 10 billion light-years away—lies an active quasar called QSO 2237+0305 that fascinates both amateur and professional observers. It's an excellent example of strong gravitational lensing and its multiple images are referred to collectively as the Einstein Cross. It turns out that you can see the Cross through a large backyard-type telescope, and the *Hubble Space Telescope* has also focused on this object. The light you see left the distant quasar more than 3 billion years before our Sun and planets formed. The massive galaxy lensing the light lies only about 400 million light-years away from us.

Types of Gravitational Lenses

There are three types of gravitational lensing:

1. **Strong lensing:** The distortions are very obvious because the light from a distant object is passing by a very massive object that is fairly close by. The light follows more than one path past the lensing object, and the observer often sees what's called an Einstein ring or even a set of multiple images of the same object all centered on the central mass doing the lensing. Most of the time the lensing object is a massive galaxy cluster.

2. **Weak lensing:** Here, the lensing object doesn't have enough gravitational pull to create rings, arcs, or multiple images. What the observer does see are often sheared images of the background object. Even from distorted, sheared images and rings, astronomers can deduce something about the background object in order to figure out what it is.

3. **Microlensing:** This is an interesting use of lensing that has led to the discovery of planets around distant stars. The lensing objects can also be stars and even stellar black holes.

Lenses and Exploration

Gravitational lenses are equal-opportunity tools for exploring the distant universe. They act on light from across the electromagnetic spectrum, so they can be used to study the last faint quivers of light from the Big Bang, called the cosmic microwave background. This is a diffuse background of light that began its journey across space some 370,000 years after the creation of the universe. It was once very energetic, hot, and possibly as bright as the surface of a star. But the expansion of the universe has stretched the wavelengths of that light, and we see it today as microwave radiation. It's faint and difficult to study. Gravitational lensing offers a way to observe changes and fluctuations in this remnant radiation that contains the last echoes of the Big Bang.

Lensing and Early Galaxies

The farther we look into the universe, the further back in time we look. In some observations, particularly with the *Hubble Space Telescope*, astronomers use gravitational lenses to see very distant galaxies as they appeared when they first began forming.

There's another clever use for gravitational lensing: the search for dark matter. This mysterious "stuff" seems to be especially thick in galaxy clusters. Not only are their own gravitational influences holding them together, but a healthy distribution of dark matter (with its own gravitational influence) is helping, too. Dark matter adds to the gravitational lensing capability of a galaxy cluster. If all galaxy clusters have significant dark matter components, then gravitational lensing becomes a very important way to figure out the distribution of this mysterious stuff throughout the universe.

THE BIG BANG

Cosmology 101

One of the most tantalizing questions we can ask in astronomy is, "How did all this material that we detect in the cosmos get started?" To answer it, astronomers look back across 13.8 billion years to a time when the universe was in a seething, hot, dense state of existence. How long it was that way and what came before it are still unknown. Yet, from that first state of existence sprang the entire universe we know today. The origin of the universe and how it has evolved to the complex and large structures we see today are the subject of a science called *cosmology*.

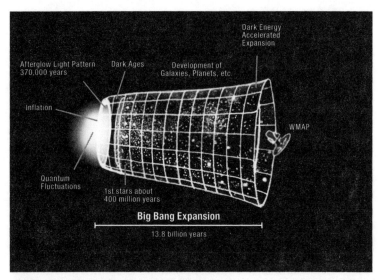

The expansion of the universe from the moment of its birth—the Big Bang—to the present day. The WMAP mission detected the faint echoes of light from the Big Bang.

The Beginning

The Big Bang—the birth of the universe—occurred some 13.8 billion years ago. It was an event that filled all of the space there was at that time with matter and energy. It was the beginning of space and time. The Big Bang wasn't an explosion, as the name seems to suggest. Rather, the birth of the universe set off an expansion of space and time that continues to this day. The most primordial particles of matter were created in the event.

The first second after the Big Bang, the entire universe was a soup of subatomic particles, superheated to 10 billion degrees. In that first second, amazing things happened:

- The force of gravity separated out from the electronuclear force and was joined soon thereafter by the electromagnetic force.
- The universe changed from being a hot soup of quarks and gluons (elementary particles), and protons and neutrons began to form.
- At the ripe old age of one second, the newborn universe was cool enough that it began forming deuterium (a form of hydrogen) and helium-3. At this point, the newborn universe had doubled in size at least ninety times!

Over the next three minutes, the infant universe continued to cool down and expand, and the creation of the first elements continued. For the next 370,000 years, the universe continued its expansion. But it was a dark place, too hot for any light to shine. There existed only a dense plasma, an opaque hot soup that blocked and scattered light. The universe was essentially a fog.

The next big change in the universe came during the era of recombination, which occurred when matter cooled enough to form atoms. The result was a transparent gas through which the original flash of light from the Big Bang could finally travel. We see that flash today as a faint, all-encompassing, distant glow called the cosmic microwave background radiation (sometimes shortened to CMB or CMBR). The universe was leaving its cosmic dark ages behind. Gas clouds condensed under their own self-gravity (possibly helped along by the gravitational influence of dark matter) to form the first stars. These stars energized (or ionized) the remaining gas around them, lighting up the universe even more. This period is called the Epoch of Reionization.

From the Big Bang to You

1. Pre–Big Bang: quantum density fluctuations
2. Pre–Big Bang: cosmic inflation
3. 13.8 billion years ago: the Big Bang
4. 13.4 billion years ago: the first stars and galaxies
5. 11 billion years ago: the Milky Way Galaxy starts to form
6. 5 billion years ago: the Sun begins to form, along with the planets
7. 3.8 million years ago: the first life appears on Earth
8. 2.3 million years ago: the first humans appear
9. Modern time: you were born

The First Galaxies

By the time the universe was nearly 400 million years old, the first stars and galaxies were forming. As the expanding cosmos cooled, dark matter condensed into clumps. This spurred the accretion of gas into dense regions, eventually forming stars. The galaxies looked nothing like the spirals and ellipticals we see today. The primordial galaxies were more like shreds of light-emitting material. They arose from fluctuations in the density of the early universe.

The universe continued to expand, but that expansion began slowing down for the next few billion years down under the gravitational influence of matter in the form of stars and galaxies. Then, when the cosmos was about 5 or 6 billion years old, something interesting happened. Dark energy—which had been around for most of the history of the cosmos—began to accelerate the expansion of the universe. This mysterious force seems to be a property of otherwise "empty" space, and its effect has been measured through observations of distant supernovae by ground-based as well as space-based observatories such as the *Hubble Space Telescope* observations of distant early supernova explosions. Dark energy acts opposite gravity and continues to speed up universal expansion today.

The First Stars

The first stars were massive, lonely objects that lived very short lives and were eventually destroyed in tremendous supernova explosions. Those catastrophic deaths resulted in the creation of black holes, which grew by sucking in gas from surrounding areas and merging with other black holes. The remains of the giant stars were blasted out into space, enriching the interstellar medium with elements heavier than hydrogen and helium. Some of those elements were created in the stars' nuclear furnaces in a process called nucleosynthesis. Others were forged during the supernova explosions through explosive nucleosynthesis.

Galaxy Evolution

From the time they formed until the present day, galaxies have been the major sites of star formation in the universe. Our own galaxy was born about 10 billion years ago, and it has grown—as many other galaxies do—through collisions and mergers. As this process has occurred, the Milky Way's regions of star formation have been busy, creating new generations of stars from the materials left behind by the first populations. Our own solar system was formed about 5 billion years ago, some 9 billion years after the first moments of the Big Bang. It will exist for another 5 billion years, until the Sun dies. Our galaxy will eventually enter into a multi-billion-year-long dance with the Andromeda Galaxy, an event that will merge the two into one large elliptical galaxy looking nothing like the spirals that combined to build it.

The universe's billions of galaxies exist in superclusters that form giant sheets and filaments of matter that stretch across the cosmos. While our universe is not going to collapse in on itself, this cosmic web of existence will continue to expand for tens of billions of years into the future.

What Came Before the Big Bang?

Nobody knows for sure exactly what preceded the creation of our universe. The answer could be that we live in one bubble of a larger multiverse. This region that contains our universe may be what formed in the Big Bang. We can't see beyond it (yet), but if we could, there would be other universes out there in their own bubbles, expanding just as ours is. Another idea is that there *was* a universe before ours, but evidence of its existence was destroyed in the Big Bang that created ours.

A SCIENCE-FICTION UNIVERSE

Wormholes in Space

Science fiction is filled with stories about faster-than-light travel and wormholes being used to cross immense distances in space. You have to admit, if you're a space traveler looking to get from one side of the galaxy to the other and don't have the 100 million years it takes to make the trip, even at the speed of light, then taking a cosmic shortcut might look pretty attractive. Wormholes in space get invoked a lot, so let's take a look at what they are and whether or not they'd make good onramps to the cosmic superhighway.

Famous Science-Fiction Wormholes

Science-fiction books, TV, and movies are filled with wormhole-users. Here are just a few examples:

- *A Wrinkle in Time* describes a method of travel across wrinkles (tesseracts) in space-time.
- Carl Sagan's *Contact* uses a wormhole-type mechanism to transport the heroine to a distant star.
- In the *Vorkosigan Saga* by Lois McMaster Bujold, the characters move through a nexus of wormholes much like a transspace subway system.
- In the movie *2001: A Space Odyssey*, one character is transported through what seems like a series of wormholes.
- In *Star Trek: Deep Space 9*, the main space station where the action takes place is situated next to a wormhole that links two quadrants of the galaxy.

Wormholes Deconstructed

Do wormholes actually exist? The truth is, nobody has actually seen a wormhole, yet. They're a theoretical construct. Essentially, the idea is that you should be able to move from one point to another in space without having to actually cross the space between the two points. Something has to "fold" that space, which would require a great deal of gravitational pull. A black hole has an incredible amount of gravitational pull, so much so that it warps space-time in its near vicinity. If it could do that, then some sort of structure could theoretically "bridge" the folds that the black hole makes.

A theoretical "fold" in space-time, with a bridge between two widely separated areas. It's possible that these could exist, but none have been found.

The idea of a wormhole bridge across the universe isn't a new one. In 1935, Albert Einstein and Nathan Rosen (1909–1995) applied their genius to the problem of crossing space and came up with what they called "bridges" linking widely separated areas of space. Some people working on the theory of wormholes suggest that these bridges could only last for a short period of time, so you'd have to really hustle to get through before the wormhole collapsed. Another problem rears its ugly head when you stop to think about what's making the wormhole possible: If you tried to use one to travel through space, you'd quickly run into the singularity—the infinitely dense heart of a black hole. That singularity would stop your trip (and you) in your tracks. You'd be sucked into the black hole and you would *not* come racing out the other side through a "white hole."

What's a Wormhole?

Physics describes a wormhole as an Einstein-Rosen bridge. One end of the bridge would be in one area of space-time, and the other end would be in another region of space-time. A wormhole can, theoretically, take you across space and also forward or backward in time.

So the question remains: Is there any way a wormhole *could* be used to travel across the light-years someday? In fact, there may be, although again only from a theoretical standpoint. You could use something called "exotic" matter to stabilize the region around the opening of a wormhole (where the black hole exists) and allow a spaceship to slip through. Provided the spacecraft is well-shielded and can withstand the conditions in and around the black hole, it could work. What kind of stuff is exotic matter? Good question. Its most important attribute is that it has to ward off the effects of gravity—a sort of gravity repellent. And that definitely *is* theoretical!

White Holes

A working wormhole includes a black hole and a white hole. Black holes are infinitely dense objects in space that have a strong gravitational pull because of the mass they contain. Material sucked into a black hole gets commingled with everything else and crushed down to its atoms. A white hole is the hypothetical reverse of a black hole: Matter and light inside can escape from it. Astronomers have yet to observe a white hole, but there's nothing in the laws of physics that says they can't exist.

Faster-Than-Light Travel

If we can't solve the problem of traveling a huge universe by hopping through wormholes, what about figuring out a way to go faster than the speed of light? In *Star Trek*, the Federation, the Klingons, the Romulans, and just about everybody else use some sort of space-warping drive to get them from A to B in short amounts of time. Again, we're looking at something that can fold space, which implies a technology far beyond anything that's feasible today. We have to remember that the speed of light *is* the top speed limit in the universe. Nothing can surpass that, even if it has infinite energy at its disposal. However, there are hypothetical particles called tachyons that always move faster than light. Science-fiction technologies invoke them a lot as tachyon drives or pulses. They could theoretically get you where you're going at a speed faster than that of light. However, you'd leave before you left, and you'd arrive at your destination earlier than you expected. So, while these things work in science fiction, there are still a few bugs to be worked out here in reality.

It turns out that the folks at NASA are working on an idea first suggested by theoretical physicist Miguel Alcubierre (1964–). He found a mathematical formula that, if applied, would envelop a ship in a warp bubble; this bubble would allow it to warp space and move faster than light. The process needs a ton of exotic matter, which engineers would have to build or find somewhere. The NASA engineers have built a prototype of their warp drive, using laser beams to create tiny warp bubbles. It's an exciting breakthrough, and perhaps in the middle-distant future, we'll see the first warp-drive ships venture "out there"—thataway.

Galactic Goof

Not all science fiction gets the details of faster-than-light travel right. For example, in *Star Wars: A New Hope*, Han Solo says to Obi-Wan Kenobi, "You've never heard of the *Millennium Falcon*? It's the ship that made the Kessel Run in less than twelve parsecs." Astronomy-savvy people in the audience howled (not necessarily out loud), since parsec is a measure of *distance* and not *time*.

EXTRATERRESTRIAL LIFE

Are *They* Out There?

These days, the possibility of finding life "out there" is an integral part of astronomy. The exploration of Mars has been spurred in large part by the search for life, or at least for conditions that could support it. The environments of some moons in our solar system—most notably Europa at Jupiter and Enceladus and Titan at Saturn—have extended biologists' Earth-based definition of suitable conditions for life to evolve. Even though these places may be colder and less hospitable to human life than here on Earth, there's still the possibility that they could support other forms, such as microbes.

Beyond the Sun, there exist countless worlds. The *Kepler* mission is on the hunt for Earth-like planets around other stars, called exoplanets, and has found many planet candidates, not all of them suitable for life as we know it. Astronomers using the European Southern Observatory in Chile have even found an Earth-sized planet circling around Alpha Centauri B, which lies 4.37 light-years from Earth. While the newly discovered planet is too hot and close to its star to be hospitable to life, the discovery is another step on the road to finding life elsewhere. The first Earth-like planet with suitable habitats for life will be found eventually, and that discovery will drastically change the way we look at the cosmos—and ourselves.

Aliens of Our Dreams

The idea that aliens have visited our planet began in the middle of the twentieth century. Sightings of unidentified flying objects (UFOs) began shortly after World War II, and those led to some amazingly complex stories about aliens crashing into the southwestern deserts of the United States or being held in freezers at secret military installations. Oddly enough, those "aliens" are almost always pictured as looking somewhat like humans, but with weird slit eyes or cat's eyes and gray or pale white skin, an almost dream-like description. Alien abduction stories (in which aliens take humans—and cows and sheep—into their ships for some nefarious purposes) often seem to mirror early science-fiction stories that covered the same territory (but in a more entertaining way). People have suggested interesting psychological aspects to these sightings and events. They often seem to occur just as the victims are going to sleep, leading skeptics to suggest that these dreams arise in the human mind when it's getting ready to shut down for the night.

The Drake Equation

There's no way of knowing right now whether or not there are aliens out there on other planets. That doesn't stop us from trying to estimate how much life could be present in the universe. Astronomer Frank Drake (1930–), who was doing radio astronomy searches for signals from alien civilizations in the early 1960s, came up with an equation that can help estimate how many civilizations could be in the galaxy. His equation looks like this:

$$N = R^* \cdot f_p \cdot n_e \cdot f_l \cdot f_i \cdot f_c \cdot L$$

where N is the number of civilizations in our galaxy that have the ability to communicate with us.

To get to N, you have to multiply the following factors:

- R^*—the average star formation in our galaxy each year
- f_p—the number of those stars that have planets
- n_e—the number of planets that could potentially support life (for each star that has planets)
- f_l—the number of those planets that actually go on to develop some kind of life
- f_i—the number of planets that actually *do* develop intelligent life
- f_c—the number of civilizations that are technologically advanced enough to advertise their existence (through radio signals, etc.)
- L—the length of time it takes for those civilizations to start releasing their "I'm here" signals

Values for most of these factors can be inserted into the equation. It turns out that even with some conservative estimates, the number of civilizations in the Milky Way alone could be nearly 200 million. Of course, this is something of a simplistic calculation that relies on some numbers for which we can only make a good estimate. It doesn't cover the civilizations that might have arisen due to contact with other, older societies that have spread their populations across the stars. But it's a good way to think about how many alien civilizations could be out there among the stars.

What Will ET Look Like?

Life on Earth arose in conditions that can be replicated across the galaxy. We came about on a terrestrial-type planet with water. It was seeded with the chemical precursors of life, which eventually formed the first primitive life forms. From there, evolutionary changes shaped what our planet's life forms looked like. Humans are the way we are—big brains, bilaterally symmetrical, capable of language, long reproductive cycle, eyes sensitive to light from the Sun, and other traits—because of specific evolutionary developments necessary for us to survive and thrive in the conditions of our planet.

There is nothing that says intelligent life that arises on another planet is going to evolve the same way we did. Let's say some folks from Tau Ceti IV drop by for a visit. They have four arms and two legs, with one giant eye and larger brains than us. Their sense of smell is attuned to their own environment, and their eyes are used to a slightly different range of light. Because they have four arms, they have a very alien sensitivity to gestures as a form of communication. What we think of as a friendly "hi" and a wave of the right hand could mean something entirely different to them. They might hear different frequencies than we do and thus not be able to suss out our speech patterns. Our skin color could even send a coded message to them that we didn't intend. It's a sure bet their language won't be like ours. So, when and if we do meet them, both civilizations will have a steep learning curve before we can communicate.

The Search for ET

People have been searching for ET by looking for radio signals from distant civilizations, and that exploration is called the Search for Extraterrestrial Intelligence (SETI). One of the most active current searches is underway through the SETI Institute in California, where they are using the Allen Telescope array of radio telescopes to listen for faint signals from other civilizations. The most likely signals could be coming from a portion of the electromagnetic spectrum called the "water hole." This is a fairly quiet part of the radio band and happens to be where neutral hydrogen gas and the hydroxyl molecule emit radio signals.

When you combine these molecules, you get water (H_2O). If water is the stuff of life throughout the cosmos, then this region of the spectrum seems like a natural place to look for signals.

One intriguing idea for finding ET is to search for the redshifted wavelengths or frequencies being emitted by their military radars, or perhaps even their radio or TV signals (if they have such things). The assumption is that if civilizations use these technologies, the frequencies used get emitted to space. Those signals would show up here at Earth as low-frequency radio emissions, which can be detected by newer radio telescope arrays. Another way to search for life elsewhere in the universe is to study starlight as it passes through the atmospheres of distant planets. Life undergoes processes that release certain chemical compounds into the air. Those atmospheric molecules absorb certain wavelengths of light, providing spectral clues to the possible existence of life. So, spectra of distant starlight could reveal whether or not life has arisen on a planet.

THE HISTORY OF ASTRONOMY

Learning about the Cosmos

The history of astronomy is the story of humans making the leap from simply observing the sky and using it as a calendar or for navigational purposes to actively exploring it and gaining an understanding of the stars, planets, and galaxies. It took more than 3,000 years to make that leap, and in the process, old worshipful attitudes toward celestial objects gave way to profound scientific interest.

How It Started

Imagine being a cave person, grubbing away with hunter-gathering tasks day after day and then hiding in your rocky home after dark so the creatures of the night don't eat you and your family. If you do venture out at night, it's for short trips not far from the cave. So one night after the bright thing in the sky has gone down below the horizon, you're just about to head back to the cave when you happen to look up and really notice the night sky in all its glory. Little bright points of light twinkle at you. Maybe there's a crescent Moon. You hardly know what to make of all this. Everything you see is beyond your reach, but you've noticed over the years you've been watching them that the lights in the sky follow the same paths year after year. Perhaps you decide to make a record of what you see—in a painting on the wall of a cave, or on an animal hide. That way you can teach others about it and add that knowledge to the information you and your clan need to survive.

Archaeoastronomy

Want to know how people in the past studied the sky? Archaeoastronomy is the application of archaeological techniques to uncover what our ancestors knew about the sky. Ancient people used the sky as a calendar and a timekeeper. In many places, they erected giant temples aligned with sky objects—such as the pyramids in Egypt, the Mayan pyramids in Central America, or the pillars of Stonehenge in England. Archaeoastronomers are most interested in the meanings that sky objects had for early cultures. This science has little to do with studying celestial objects and more to do with learning how ancient people entwined what they saw in the sky with their cultural beliefs and practices.

The Sky as a Place of Worship

Every culture on our planet has watched the sky and wondered about the complex motions of planets, the Moon, the Sun, and the stars throughout the year. For a long time, these objects were considered gods and goddesses, and people worshipped them. The ancient Greeks, Egyptians, Chinese, Hindus, Aztecs, Maya, Koreans, First Nations, and many other people thought of the sky as a dwelling place for their deities. Their astronomy was exclusively naked-eye observing, and they used the Sun, Moon, and stars for navigation and timekeeping.

The Scientific Sky

The first scientifically minded celestial observers included people such as Nicolaus Copernicus (1473–1543), Johannes Kepler, and Galileo Galilei, who began looking at the sky through telescopes they built. Galileo's view of Jupiter in 1610 transformed our view of the planets. They weren't just dots of light in the sky. They were worlds.

Over the years, more and better telescopes have revealed double stars and nebulae in the sky, and their discoverers set out to figure out what these things were. The science of "natural philosophy" uses mathematics, chemistry, and physics to explain objects and events in the universe. Nicolaus Copernicus came up with the heliocentric solar system, with the planets orbiting the Sun. The laws of planetary motion developed by Johannes Kepler and the laws of physics devised by Sir Isaac Newton helped explain the motions of bodies in space.

The Laws of the Universe

Astronomers understand and apply the laws of physics and orbital motions in such widely varied areas of research as exoplanet discovery, the search for dark matter in and around galaxies, and the actions of material around black holes.

Advances in astronomy came as quickly as telescopes and instruments could be built. The science of astrophysics grew as methods in photography and spectroscopy were applied to astronomical observations. Hooking up a camera to a telescope allowed astronomers to capture images of dim and distant objects by taking time exposures. In 1876, Vega was the first star ever photographed.

Passing the light from celestial bodies through spectrographs gave astronomers insight into the chemical makeup and physical processes that governed stars, planets, nebulae, and galaxies. In the late 1800s and early 1900s, a new breed of astronomical observatories popped up around the globe, equipped with multi-wavelength sensors and specialized cameras designed to capture as much of the distant universe as possible.

Today's astronomers and astrophysicists study the sky with arrays of ground-based and space-based instruments that would amaze and delight the sky observers of old. Imagine what Kepler would say if he knew his name was on a probe searching out planets orbiting other stars. Or what Galileo would do to get his hands on data from the Jupiter mission that bore his name. Edwin Hubble, the man who discovered that the universe is expanding, would be delighted to find out just how large it really is.

Omar Khayyám and the Calendar

There is a gap in Western astronomical research that coincides with the Dark Ages in Europe (roughly 400–800) and the rise of the Islamic Empire that covered much of the Middle East, North Africa, and Spain. From the eighth through the fourteenth centuries, Islamic astronomers focused their attention on translating and preserving the ancient Greek writings about astronomy. They also developed mathematical tools to tackle problems in timekeeping and calendar making and to create accurate tables of the sky and its objects. Even today, many stars in the sky bear Arabic-based names. One of the best known of these Islamic scholars was Omar Khayyám (1048–1131), a mathematician, astronomer, and poet. He built an observatory at Isfahan in ancient Persia, now Iran, and used his observations to create a calendar that is still used in Iran and Afghanistan. Eventually, with the flowering of the Renaissance, European scholars took interest in astronomy, and Khayyám and his fellow scholars passed along their knowledge to them.

NICOLAUS COPERNICUS
The Copernican Revolution

There are more than 11,000 working astronomers in the world today, and each of them stands on the shoulders of great thinkers of the past who saw the sky as something to explore scientifically. Of the many "fathers" of modern astronomy that people point to today, the first major contributions to our understanding of the stars and planets came from a man who did his work without a telescope. From his calculations came a new way to look at the universe. His name was Nicolas Copernicus, and he lived during the European Renaissance.

Copernicus did his work during a time of great cultural and scientific advances that began in Florence, Italy, in the fifteenth century and lasted into the seventeenth century. In Florence, politics, art, religion, and a population of people interested in expanding their education came together. Their ideas spread like wildfire throughout Europe due to the invention of the printing press. In time, the ideals of the Renaissance revolutionized not only art and music, but also science and technology. They even challenged the unchanging views of the Church, particularly in astronomy and the natural sciences. The discovery of America in 1492 further changed people's worldview and sustained an intellectual environment that encouraged questioning of old ideas and exploration of new ones.

The Earth-Centered Universe

Before the advances of the Renaissance, classical astronomy was still very much influenced by the Greek astronomer and philosopher Aristotle (384 B.C.–322 B.C.), who observed and charted the sky. He suggested that Earth was the center of the universe. The Sun and planets orbited our planet, and the stars were fixed on a sphere and never moved. This was all very nice, but it didn't jibe with observations. The Sun, Moon, and particularly the planets, did *not* move the way Aristotle thought they should. Specifically, some planets seemed to reverse course in the sky. To explain that, observers such as Claudius Ptolemy (90–168) charted those same motions and devised artificial little circles called *epicycles* as part of a larger orbit to explain these puzzling planetary motions.

Although subsequent discoveries showed it to be wrong, the Ptolemaic system sort of worked, but eventually tables of the Sun, Moon, and planet positions in the sky based on its calculations grew hopelessly out of date. New observations and calculations were needed to update planetary position charts. That was the situation in the Renaissance, when Western scientists once again took up the problem of predicting celestial motions. The newly invented telescope, first put to celestial use by Galileo Galilei, helped. But it was the theoretical work of Nicolaus Copernicus that set in motion a revolution in scientific thought.

Copernicus's Legacy

It may seem obvious to us today, but in Copernicus's time, the idea that the planets orbited the Sun was heresy. Once his book was published, however, the idea that the Sun was the center of the solar system spurred others—most notably Galileo Galilei, Johannes Kepler, and Isaac Newton—to base their work upon his.

Copernicus, the Renaissance Man

Nicolaus Copernicus was the son of a wealthy businessman. He studied mathematics and astronomy at the University of Cracow in Poland and went on to study medicine and law in Italy. It was during his time in Italy that he began questioning the Ptolemaic explanation of planetary motions. He became intrigued with finding mathematical calculations that would explain the motion of planets and other celestial bodies.

With his education completed, Copernicus returned to Poland to live with his uncle, who was a bishop in the Catholic Church. Eventually, he became a church canon and also practiced medicine. But the astronomical problems that had intrigued Copernicus in college continued to fascinate him, and he kept working away on them. In his later years, he also worked as an economist, a translator, and an artist, as well as serving as a diplomat.

Through his observations and work, Copernicus challenged the old idea of Earth being the center of the universe and suggested that the Sun is the object around which all things orbit. He showed, with mathematical precision, how the solar system's bodies were arranged in orbits. He pointed out that instead of being fixed in space, Earth itself follows an orbit around the Sun, as do the other planets.

His ideas were radical at a time when some scientific advances weren't well-received by church and civil authorities. The Church held to the classical Greek ideal that Earth—and its human population—was the center of creation. Putting the Sun at the center of the solar system was a demotion for Earth, and that went against the teachings of the Church.

Nonetheless, Copernicus went on to write his famous book *De revolutionibus orbium caelestium* (*On the Revolutions of Heavenly Spheres*), which outlined in great detail his heliocentric model of the heavens. It was published just two months before he died in 1543. Many scholars point to this book as a first important step into modern astronomy, and consider it the beginning of what's now called the *Copernican Revolution*. Without his work, modern astronomy would not be what it is today.

Copernicus on Money

In addition to his work in mathematics and astronomy, Nicolaus Copernicus was also an economist. He wrote treatises on the monetary exchange rates and the economic values of goods and services in a society. His interest in these subjects stemmed directly from his work as an administrator and canon in the Church.

GALILEO GALILEI

Astronomy's First Maverick

One of the great discoveries in astronomy took place outside Padua, Italy, on the evening of January 7, 1610. Astronomer Galileo Galilei aimed his newly built telescope at the planet Jupiter. What he saw astounded him. Jupiter was not just one body. He described three fixed stars appearing in his eyepiece. Galileo made a drawing of what he saw, and over the next few nights he watched in amazement as one more point appeared. All four of those "fixed stars" seemed to change position with respect to Jupiter. It was an observation that changed the way astronomers view the cosmos because here was a planet that had its own moons circling around it just as planets orbit the Sun. Suddenly our position in the cosmos was not at the center of things.

The Musician's Wily and Rebellious Son

Galileo Galilei was born in 1564, during the later years of the Renaissance, a time when the arts and sciences, politics and religion were undergoing change and rebirth. His father was the accomplished composer and music theoretician Vincenzo Galilei. Vincenzo published books on tuning systems and vocal counterpoint that went against the established musical ideas of the day. He experimented with musical styles that were edgy for his time but that we accept today as being part of the Baroque style of music. With such a freethinker as Vincenzo for his father, it's no surprise that Galileo also rebelled against the establishment—and not just in his astronomical work. Galileo never married but did have time to father two daughters and a son with a woman named Marina Gamba. The daughters eventually joined convents, and the son was later legitimized as Galileo's heir.

The Father of Modern Observational Astronomy

As a university student, Galileo studied medicine but was drawn to mathematics and physics. Eventually, he left medicine behind and took up what was called "natural philosophy"—the study of nature and the physical underpinnings of the universe. At the same time, he was studying what we now call "design" and ended up teaching art in Florence, Italy. He also gained an appointment as chair of mathematics in Pisa and began teaching geometry, mechanics, and astronomy. In many ways, Galileo was the embodiment of a "Renaissance Man"—well versed in the arts, engineering, and the sciences.

As a scientist, he was an experimentalist. In one famous test, he suggested a way to challenge the Aristotelian idea that heavy objects fall faster than light ones. Climb the Leaning Tower of Pisa, he said, and drop two balls of different masses to the ground. If they hit the ground at the same time, it disproves Aristotle's contention. It's not clear when or if Galileo actually did the experiment, but in his book called *On Motion*, he predicted that all masses are acted on equally by the acceleration of gravity and that two balls of different weights would both hit the ground at the same time.

Did You Know?

Galileo Galilei did not invent the telescope. That honor goes to two Dutch-German opticians, Hans Lippershey and Zacharias Janssen, who were master lens grinders, and Jacob Metius, an instrument maker. Their model inspired many others to try building telescopes and spyglasses for use at sea. When Galileo Galilei saw the design, he built his own telescope, which he used to make the first observations of Jupiter, the phases of Venus, sunspots, and the stars.

Galileo encountered and embraced the heliocentric theory first conceived by Nicolaus Copernicus a half-century before his birth. He defended the idea that the Earth and planets revolve around the Sun and set out to prove it. He had to take on the Church and other observers who supported the old Aristotelian view of the universe. In his writings and experiments, Galileo attempted to show that science and scripture could coexist. Still, Church authorities wanted to ban any Copernican teachings, which did not sit well with Galileo. Although he was warned not to discuss the heliocentric theory publicly, Galileo decided to write about it anyway. His major work, *Dialogue Concerning the Two Chief World Systems*, used the artifice of a series of discussions between philosophers and a layman. One philosopher represents the Copernican ideas, while the other supports the ideas of Aristotle and Ptolemy. The layman is an impartial observer.

Church officials were outraged when the book came out and threatened him with imprisonment for promoting ideas against Church teaching. He continued to write about what he observed, and in 1633, he was convicted of heresy and ordered to recant his ideas. His book was put on the Church's list of forbidden works, and authorities ordered that no one publish any of Galileo's writing. He was put under house arrest for the rest of his life. This did not stop Galileo from observing the sky and writing about his work, which he continued to do until his death in 1642. His last book, *Discourses and Mathematical Demonstrations Relating to Two New Sciences*, published in 1638, covers many of the scientific experiments and research that Galileo did. It was written during his house arrest and published in the Netherlands (which did not care about the papal edict concerning Galileo's publications).

Galileo's Contributions to Astronomy

Galileo's astronomical accomplishments comprise a substantial legacy of observational science. He was the first to aim a telescope at the heavens, and with it he discovered the Galilean satellites of Jupiter, observed the phases of Venus, and made the first detailed observations of sunspots—regions on the Sun that we know today are connected to solar magnetic activity.

Galileo and the Church

The Catholic Church quietly dropped its prohibition against Galileo's writings in 1758, but it took another 242 years for the Church to recant its own treatment of Galileo. In 1992, Pope John Paul II finally admitted that the Galileo affair had been "badly handled."

JOHANNES KEPLER
Mathematics Prodigy and Astronomer

When Johannes Kepler was a small child, he witnessed an unforgettable sight—the Great Comet of 1577. Because its passage was so close to Earth, it appeared large and bright in the nighttime skies. In another part of Europe, Tycho Brahe (1546–1601) a Danish nobleman with a passion for astronomy, made careful measurements of the comet's position. Years later, in an odd twist of fate, Kepler came to work for Brahe, and the elder astronomer's notes about the comet's passage and the motions of planets across the sky encouraged Kepler to formulate the three laws of planetary motion that bear his name.

Carrying the Heliocentric Torch

Johannes Kepler was born in 1571 in Germany. As a child he developed a love for mathematics, and by the time he was six, he was hooked on astronomy. In college he studied philosophy and theology and learned about the Earth-centered cosmology of Claudius Ptolemy and the Sun-centered solar system that Nicolaus Copernicus had developed. The heliocentric idea appealed to Kepler, but in those days, the idea that Earth was revolving around the Sun was considered heretical. In the early 1600s, Galileo Galilei had been placed under house arrest for even suggesting the idea. Yet it was clear to Kepler that the orbits of planets around the Sun helped explain their motions across the starry night skies.

In 1594, at the age of twenty-three, Kepler became a math and astronomy teacher at what later became the University of Graz in Austria. He immediately began working on a defense of the Copernican system. He was also fascinated with the idea that the universe was based on harmony and geometry. He did a great many calculations, trying to make the observed orbital positions of planets fit inside a nested set of geometrical solids that would yield the distances of the planets from the Sun. This, he felt, would help reveal God's plan in creating the universe with perfect geometry. This is not surprising since Kepler had earlier trained for the ministry and was a very devout man. He felt that science and religion could coexist and that the universe was really made in God's image. He outlined his thoughts about this and defended the Copernican heliocentric ideas in a book called *Mysterium Cosmographicum*, (Latin for *The Cosmographic Mystery*). It was published in 1596 and immediately identified Kepler as one of the leading astronomers in Europe.

The *Kepler* Mission

NASA has a spacecraft named after Johannes Kepler that searches out planets around other worlds. Launched in 2009, it looks for the periodic dimming of light from distant stars that might indicate the existence of planets there. Astronomers use *Kepler* data to calculate orbits for any planets found by the spacecraft using Kepler's laws of orbital motion.

The Tycho Brahe Years

After the publication of *Mysterium Cosmographicum*, Kepler began a correspondence with Brahe, an older and more experienced astronomer who was building an observatory near Prague. Brahe was well known as a skilled observational astronomer. In his letters to Kepler, he criticized some of his younger colleague's work, warning him that he was relying on inaccurate data. Kepler accepted an invitation to visit with Brahe, hoping that the older man might be able to provide him some support. The timing of the invitation was fortuitous: Things were not going well in Graz—the Church wanted Kepler and his family to convert to Catholicism in order to keep his job. He was not interested in converting, so he pursued work with Brahe. Eventually, the family moved to Prague.

The next few years were filled with Kepler's analysis of the older man's extensive planetary and stellar observations. In 1601, Brahe died, and Kepler was appointed imperial mathematician in his place. He then spent the next decade working on a set of star catalogues and planetary charts called the Rudolfine Tables, named for the Holy Roman Emperor Rudolf II. The tables were published in 1627 and dedicated to Emperor Ferdinand II.

Imperial Mathematician/Imperial Astrologer

One of Kepler's jobs in his role as advisor to Emperor Rudolf was to provide astrology readings, something he had done as a college student and was very proficient at. As a court advisor, his horoscopes were much in demand, even though he felt they had no scientific basis for predicting the future. But the job allowed him to continue his work analyzing Tycho Brahe's vast sets of observations. In 1609 Kepler published his first two laws of planetary motion. They were not well received because at the time most astronomers believed that orbits had to be in perfect circles. However, precise calculations of Mars's orbit based on years of observational data convinced Kepler that planetary orbits were elliptical. It took another decade of mathematical calculations before he had enough insight to develop his third law of planetary motion.

Kepler's Laws of Planetary Motion

1. Kepler applied Brahe's observations in an effort to solve the problem of Mars's peculiar-looking orbit. He reasoned that Mars

followed an elliptically shaped orbit. This led to his first law of planetary motion: *A planet moves in an ellipse around the Sun and the Sun is at one focus of the ellipse.*

2. By applying some geometry to the problem, Kepler then devised his second law of orbital motion: *A line connected between a planet and the Sun sweeps out equal areas in equal times as the planet orbits the Sun.*

3. Kepler then applied his mathematics to calculate orbital periods (that is, how long a planet takes to go around the Sun). His third law is the result: *The square of the period of a planet's orbit is proportional to the cube of the semimajor axis of that orbit.*

Johannes Kepler died in 1630, leaving behind a legacy of research in astronomy, mathematics, and philosophy of science that still intrigues modern scientists. Along with Nicolaus Copernicus and Galileo Galilei, he is credited as one of the early fathers of modern astronomy.

Kepler's Publications

Johannes Kepler published treatises about many topics. Here is a list of some of his other astronomy-related works.

1. *Astronomia Pars Optica* (*Optics in Astronomy*)

2. *Astronomia Nova* (*The New Astronomy*)

3. *Dissertatio cum Nuncio Sidereo* (*Conversation with the Starry Messenger*, an endorsement of Galileo Galilei's observations)

4. *Harmonice Mundi* (*The Harmony of the Worlds*, in which Kepler describes harmony and congruence in geometry and presents his third law of planetary motion)

THE HERSCHELS

A Family Affair with Astronomy

All of us know about families in which the parents are smart and the kids get good grades and do great extracurricular activities. The Herschels of England were that kind of family, starting with Sir Frederick William Herschel (1738–1822), his sister Caroline (1750–1848), and his son John (1792–1871). Between them, this talented trio of German-English stargazers were responsible for:

- Discovering one planet
- Charting hundreds of double and triple star systems
- Discovering planetary nebulae
- Completing deep-sky surveys of distant non-stellar objects including nebulae, galaxies, and comets
- Discovering infrared light
- Creating massive catalogs of sky objects still in use to this day

As if that weren't enough, the trio built telescopes and even explored the world of the very small by examining plants and animals under the microscope. Their work is enshrined in the Herschel Museum of Astronomy in Bath, England.

The Clan Patriarch

Sir Frederick William Herschel was born in Hannover, Germany, in 1738. He trained as a musician and eventually composed twenty-four symphonies. He played violin, cello, oboe, harpsichord, and the organ. He moved to England, where he was first violin for the Newcastle orchestra. From there, he moved to Leeds, where he took a job as a church organist and ended up becoming director of the Bath orchestra. His love of music eventually led him to explore mathematics and the study of optics. He made friends with the Royal Astronomer, Sir Nevil Maskelyne, and began building telescopes.

What's in a Name?

Sir William Herschel came up with the word *asteroid* to describe tiny worlds in the smaller solar system. He also deduced that the solar system is moving through space after he noticed that stars have *proper motion* (that is, they also have motion through space). His telescope was powerful enough that he could make out the Martian ice caps through it and he observed that they grew and shrank as the seasons changed.

In 1773 Herschel commenced recording numerous observations of the stars and planets. He kept careful observational journals all his life, often going back and re-observing double stars and other distant objects to note any changes in their appearances or positions. His notes formed the basis of his famous Deep Sky Catalog and Double Star Catalog, as well as his *Catalogue of 500 New Nebulae, Nebulous Stars, Planetary Nebulae, and Clusters of Stars*. Using his observations of the stars over the years, Sir William Herschel was also able to suggest that the Milky Way Galaxy had a disk-shaped structure.

Discovering Uranus

Sir William Herschel spent much of his time searching for double stars and ultimately found nearly a thousand multiple-star systems.

As he was observing in March 1781, he thought he spotted something that looked distinctly non-stellar. He marked it down as a possible star or comet and moved on. Further observations showed that it was moving, although very, very slowly. Herschel worked with a Russian scientist named Anders Lexell (1740–1784), who calculated an orbit for this mysterious object and suggested that Herschel's discovery could be a planet. Herschel looked at the data and agreed. He named his find the "Georgian star" after King George III of England, but the convention was to name planets after classical mythological features. So, Herschel's discovery became known as Uranus after the Greek god of the heavens.

Herschel and Infrared Light

Sir William Herschel's interest in observing the sky led him to study the Sun. However, since even a small glance at the Sun can cause eye damage, he investigated various ways to filter sunlight so that he could safely observe sunspots—those regions on the Sun that are cooler than their surroundings and are involved in solar activity. Herschel's experiments with a red filter yielded an interesting result: He noticed that even though he didn't see any light passing through the filter, he could feel heat. He used a thermometer and found the temperature of this "invisible" light to be quite warm. This unseen light was beyond the red end of the visible spectrum, and so it became known as *infrared*.

Caroline Herschel

At a time when women were not expected to have much interest in science, Caroline Herschel was as intrigued by the night sky as her famous brother, Sir William Herschel. She was born in 1750 and grew up a sickly child. When she was seventeen, she left the family home and moved in with William, who had emigrated to England. Like her brother, she studied music and became an accomplished vocalist. Also like William, Caroline was bitten by the astronomy bug. She worked with him as he built ever better telescopes, lending her hand at mirror polishing and mounting the telescope hardware. Her brother taught her how to reduce data, and she began doing her own observations. She discovered eight comets, and eventually the government recognized her services, particularly in support of Sir William's research, by paying her for her work. She may well be the first woman in modern time to be paid to do astronomy, and she spent much of her life re-observing and verifying her brother's original astronomical targets.

Sir John Frederick William Herschel

Like his father and aunt, Sir John Herschel was interested in mathematics, astronomy, botany, and chemistry. He became interested in astronomy in 1816 and built his first telescope. It wasn't long before he was following in his father's footsteps as an observational astronomer. He re-observed many of the multiple star systems catalogued by his father and devised the Julian day calendar system. In 1833, Sir John and his wife traveled to South Africa, a trip that allowed him to complete his father's survey of stars and nebulae. He published *The General Catalog of 10,300 Multiple and Double Stars*, as well as the *New General Catalog of Nebulae and Clusters* (which is referred to today as the NGC).

The *Herschel* Mission

The European Space Agency named its far infrared and sub-millimeter orbiting telescope after Sir William and Caroline Herschel. The mission is making its name in the study of star formation and is part of ESA's science program of four satellites, including *Rosetta*, *Planck*, and *Gaia*.

ISAAC NEWTON

The Man Who Transformed Physics and Astronomy

We live in an expanding cosmos in which objects are constantly in motion. Many of the laws that describe these motions were devised by an Englishman who started out life as a farmer but ended up being one of the most influential scientists of all time. His name was Sir Isaac Newton, and his laws of orbital motion and gravitation are fundamental rules that every astronomer learns. Farming's loss was science's gain.

Newton's Three Laws of Motion

Isaac Newton carefully worked out the math describing an object's motion in space. On the basis of this, he postulated three laws:

1. An object at rest will stay at rest. If it's in motion, it will stay in motion unless it's acted upon by another object or force.
2. Acceleration occurs when a force acts on a mass. The greater the mass, the more force is needed to accelerate the object. Physics and astronomy students learn this as $F = ma$, where m stands for the amount of mass and a stands for acceleration. In the case of an apple falling to the ground, you plug in the mass of the apple and its acceleration, and that gives you the force (in units called, appropriately, Newtons).
3. For every action there is an equal and opposite reaction. If one ball hits another ball in a bowling alley, the first ball is pushed back in the opposite direction as hard as the second ball is pushed forward.

These may seem like obvious rules to us today, but in Newton's time they were amazingly new ideas.

Newton's Law of Gravity

One of the most important rules of nature that Newton developed was his universal law of gravitation. This law says gravity is a force that acts on all objects in the universe. Newton pointed out that you can calculate the force of gravity between two objects if you know their mass and their distance apart. The closer together the objects are, the stronger the force of gravity that attracts them to each other. The farther apart they are, the weaker the gravitational pull. Furthermore, objects orbiting a strong center of gravity will orbit more slowly if they're farther away and faster if they're close. Gravity influences everything from star formation, the creation of black holes, galaxy evolution, and orbital mechanics of objects in our solar system right down to the interactions between atoms.

Newton and Orbital Motion

Orbital motion is an important part of life in the universe, so let's take a closer look at it. Orbits illustrate Newton's First Law of Motion, which states that an object in motion will stay that way unless something acts on it to change the motion. You experience this every day. You see a door about to slam and use your hand to stop the action. Your hand is the force that changes the door's speed and direction.

The force that changes motion in space is gravity. Physical bodies all have mass. Each mass has an attractive force on other masses. The more massive the body, the stronger the gravitational force. Gravity and velocity keep planets in orbit around the Sun, moons going around planets, and galaxies orbiting other galaxies.

Newton's Life and Work

Isaac Newton was born in 1642, the same year that Galileo Galilei died. His student days were lackluster, and although he was raised to be a farmer, his talents lay elsewhere. He went to Trinity College, Cambridge, to receive a classical education in logic, ethics, physics, and the works of Aristotle. He left after graduation and spent the next several years working on mathematics. He came up with the basic principles of differential calculus at the same time that, coincidentally, those principles were being established in Germany by Gottfried Leibniz (1646–1716). He also dabbled in optics and spent time working out a mathematical theory of circular motion. By 1667 he was a fellow of Trinity College and was well known for his mathematical ability. He continued to study optics and began working on a telescope of his own design, which used a reflecting mirror instead of a lens. Similar types of telescopes are known today as Newtonian reflectors.

Based on his continuing work in math and optics, Newton was made Lucasian Professor of Mathematics in 1669 (the same chair held today by Stephen Hawking). He published tracts about optics, light, and color and eventually turned his attention to celestial mechanics. His interest was piqued by an exchange of correspondence with Robert Hooke (1635–1703), a fellow scientist. Although offended by some of Hooke's criticisms of his work, Newton was intrigued enough by the correspondence—plus the appearance of a comet beginning in 1680—to work out the details of planetary orbits. He wrote proofs of his work and published his three universal laws of motion in a book called *Principia Mathematica*. Its publication made him an international star in scientific circles, and his work on describing motion contributed greatly to the industrial revolution that began during his lifetime.

Newton to the Stars!

The European Space Agency named its *XMM-Newton* x-ray astronomy mission after Sir Isaac Newton and his many accomplishments.

Sir Isaac Newton's later life saw him thinking deeply and writing about religious and metaphysical matters. He worked as warden of the Royal Mint for thirty years, beginning in 1696. He also served as President of the Royal Society and was knighted in 1705 by Queen Anne. He died in 1727, leaving behind a tremendous scientific legacy.

The Case of the Falling Apple

Every school child who learns about Newton is told the story of how he came up with the theory of universal gravitation: An apple fell out of a tree and hit him on the head. There's no proof that Newton was conked on the head, but he did say that he watched an apple fall and wondered why it fell straight down and didn't wander sideways. He eventually concluded that the gravitational pull of Earth was somehow proportional to the gravitational effect of the apple on Earth.

HENRIETTA SWAN LEAVITT

The Human Computer

On March 3, 1912, the Harvard College Observatory released one of its astronomy circulars announcing measurements of the periods of variable stars in the Small Magellanic Cloud. It was a routine announcement about stellar brightnesses in various parts of the sky, but it also contained an important statement: "A remarkable relation between the brightness of these variables and the length of their periods will be noticed."

That sentence proved a warning shot across the bow for astronomers arguing that distant galaxies were part of the Milky Way. More importantly, that simple statement is the first notice of the remarkable life's work done by Henrietta Swan Leavitt (1868–1921)—a woman who loved astronomy and worked in astronomical research but, because of her gender, was actively discouraged by her boss from doing theoretical work. Yet she went on to make a major discovery in astronomy, one that enables others to determine accurate distances to distant objects.

Work on Cepheid Variables

Henrietta Swan Leavitt was the daughter of a Massachusetts church deacon and attended Oberlin College and what became Radcliffe College in Cambridge, Massachusetts. She fell in love with astronomy in her senior year and eventually volunteered at the Harvard College Observatory under director Charles Pickering. She became a paid assistant, doing whatever work he assigned to her. Pickering assigned her to study variables, stars that vary in brightness over predictable periods of time. She picked out thousands of these flickering objects on the observatory's photographic plates and published a paper in 1908 noting that bright variables appeared to have longer periods.

Cepheid Variables

There are many variable stars in the sky. Our own Sun is considered a variable since its brightness rises and falls over time. There's a special class of variables called Cepheids, named after a star found in the constellation Cepheus. These vary in brightness according to a regular pattern. Cepheids have masses between five and ten times that of the Sun. At their brightest they contract a bit and become very bright. Then, pressures build up inside them, and they are forced to expand, which makes them appear dimmer. Then they shrink in again, and the cycle repeats itself. The period of this pulsation is directly related to the intrinsic brightness of the star, and astronomers use that period to determine how far away it lies. If it lies in a distant galaxy, then its distance is the same as its galaxy. So, determining how far away a Cepheid variable is tells us how far away its galaxy is. The connection between a star's luminosity and its variation in brightness became known as the period-luminosity relationship.

Over time, Ms. Leavitt discovered more than 2,400 variable stars and also found four novas. Yet her most important contribution was her work on the Cepheids. It allowed Edwin Hubble to measure the distance to one in the Andromeda Galaxy in 1923 and thus prove that galaxies lie outside the Milky Way. Without her work, he might not have made his discoveries, and he gave credit to her, noting that she should have been given the Nobel Prize for her own momentous finding.

The Science of Photometry

Astronomical measurements mostly focus on the light from distant objects. Photometry is the science of measuring the actual intensity of electromagnetic radiation emitted or reflected by something in the sky. In the past, the intensity of light from an object was measured with special instruments called photometers. These instruments revealed something about the object's intrinsic brightness. Today these detectors incorporate highly sensitive sensors called charge coupled devices (CCDs) to detect and very accurately measure the intensity of light from distant objects. In Henrietta Leavitt's day, photometry was performed by passing light through the photographic plates of starfields and measuring the light that made it through the plate.

In 1921, due to her painstaking work, Henrietta Leavitt was appointed head of stellar photometry at the Harvard Observatory. Unfortunately, she was suffering from cancer and died later that year. She was one of the first female astronomers to be recognized for major contributions to the field, and her legacy lives on today as scientists use her work to establish accurate distances of faraway celestial objects.

Women in Astronomy

The status of women in astronomy has improved since the days when Henrietta Leavitt and her coworkers Annie Jump Cannon, Williamina Fleming, and Antonia Maury were degradingly referred to as "Pickering's Harem" while they worked at Harvard College University. It was fairly common in the early years of the twentieth century for male colleagues to discourage women from doing astronomy. In some cases, they were refused the use of telescopes reserved exclusively for men. Today, there are many women doing astronomical research, and astronomy professors report about equal numbers of females and males taking astronomy classes. However, there are still fewer women getting graduate degrees in astrophysics than men, and the percentage of women heading departments of astronomy and astrophysics remains very low. Various organizations, including the American Astronomical Society, are actively working to encourage interested women to study and work in astronomy.

EDWIN P. HUBBLE
Cosmological Revolutionary

Most people know of Edwin P. Hubble because of the space observatory that bears his name. The *Hubble Space Telescope* is one of the great workhorses of astronomy and has shown astronomers fascinating objects and processes throughout the universe. Its name honors one of the astronomical stalwarts of the early twentieth century.

What did a former lawyer from Marshfield, Missouri, do to get his name on a telescope? He proved that some distant clouds of light in the sky are actually distant galaxies that lie well beyond the boundaries of our own home galaxy, the Milky Way. In a very real sense, his work presaged the great work of the telescope that bears his name.

From Law to Astronomy

Edwin P. Hubble was born in 1889, the son of an insurance executive. He was interested in science as a child, studied mathematics and astronomy in college, and at his father's urging, studied law at Oxford University in England. When he turned twenty-five, Hubble decided to make a profession of astronomy and attended the University of Chicago, earning his PhD in 1917. His major interest was the study of so-called faint nebulae. After serving in the army in World War I, Hubble joined the staff of Mount Wilson Observatory in California, where he used the newly finished 100-inch Hooker telescope to study these objects in more detail.

Hubble's Major Accomplishments

Edwin Hubble put his access to one of the world's best telescopes to good use. While studying the Andromeda Spiral Nebula (today known as a spiral galaxy) in 1923, he discovered the flicker of a Cepheid variable star. These are used as "standard candles" in determining distances in the universe. His discovery answered a question: whether or not these spiral nebulae were inside our own galaxy or whether they lay much farther away. Using his measurements of Cepheids, Hubble was able to show that they were very distant and definitely not part of our own galaxy. Up until that time, many astronomers held the view that the Milky Way was the entire cosmos. Hubble's discovery showed, for the first time, that the universe was much larger than anyone thought. It was a revolutionary finding. Astronomers continue to use Cepheid variables as one part of a cosmic distance toolkit to determine how far away objects are and how fast they are moving. This remains one of Hubble's greatest contributions to astronomy.

Hubble also discovered that objects in the universe seem to be moving apart, thus showing that the universe is expanding. He determined that the velocity of this so-called "recession" is faster the farther away an object lies from us. This idea of an expanding universe rocked astronomy and is one of the principal foundations of the science of cosmology. Others had suggested that the universe might be expanding; Hubble went to work to calculate a rate of expansion based on his observations. That expansion rate came to be called the Hubble Constant, often noted in astronomy literature as H_o (pronounced "H-naught"). Hubble originally calculated it to be about 500 kilometers per second per megaparsec. Today, with more sensitive telescopes and techniques, the value of H_o has been adjusted to 67.15 ± 1.2 kilometers per second per megaparsec.

Hubble's Law

How did Hubble figure out the recession velocities of galaxies? He used something called the *Doppler Effect*. This says that waves of light or sound have a higher frequency (if they're sound) or higher wavelength (if they're light) as they move toward an observer and lower frequency/wavelength if they are moving away. He used a spectroscope to split the light from distant galaxies into a spectrum and then noted that the spectra of many galaxies were shifted lower—toward the red end of the electromagnetic spectrum. Galaxies moving toward Earth had spectra that were shifted toward the blue end, and are referred to as "blueshifted."

Edwin Hubble observed many galaxies in his career, and he set to work classifying these objects by their shapes. His *Hubble sequence of galaxy morphologies* is the basis for classifications still used today. Galaxies can be spiral, elliptical, lenticular, or irregular. In modern astronomy, the classifications of spirals in particular are subdivided into spirals with tightly wound arms and big central bulges, spirals not so tightly wound and having fainter bulges, and loosely wound galaxies with very faint central regions. Lenticulars have bright central regions but no spiral arms and look similar to the elliptical galaxies (which also have no spiral arms). Irregulars are just that— blob-shaped galaxies with no spiral arms, but often showing brilliant regions of star formation. The Large and Small Magellanic Clouds are good examples of irregulars.

Edwin P. Hubble continued his work at Mount Wilson until his death in 1953. His work revolutionized astronomy and cosmology, and it's no wonder that the *Hubble Space Telescope* is named for his life and accomplishments.

The Great Debate

Before Edwin Hubble's work with Cepheids, astronomers disagreed about whether galaxies were "island universes"—a term first used by philosopher Immanuel Kant (1724–1804) to refer to distant nebulae that could be outside the Milky Way—or if they were simply part of the Milky Way. Astronomer Heber Curtis (1872–1942) was an island-universe guy. He observed a nova in the Andromeda Galaxy in 1917 and used its light to measure the distance to this neighboring galaxy. Although his measurement was not quite accurate, it was enough to make the case for Andromeda being a distant galaxy. Others disagreed with that hypothesis, and so a great public debate was held in 1920. The great Harlow Shapley (1885–1972) pointed out (among other things) that the Milky Way was the whole of the universe and said that a recent nova in Andromeda had been brighter than the galaxy's core. He contended that this brightness meant that the nova was very close and part of our own galaxy. Thus, Andromeda (of which it was part) was itself part of the Milky Way. Edwin Hubble's discovery of Cepheids in Andromeda resolved the debate in 1925, and he used distance calculations for those variable stars to prove that Andromeda and other galaxies with Cepheids were outside the Milky Way.

ALBERT EINSTEIN

A Twentieth-Century Genius

Albert Einstein was one of the greatest scientific thinkers of all time. His special insights in theoretical physics also led to advances in understanding in astronomy and quantum mechanics (the physics of the very small). His most famous equation is $E=mc^2$, which established that if you take the mass of an object and multiply that by the speed of light squared, you get its energy content. This concept is particularly important in astronomy and astrophysics, since it helps explain the way that stars convert mass into energy and provide light and heat. In 1921 Einstein received the Nobel Prize for his contributions to physics; the Nobel committee cited his work on mass and energy as well as his discovery of the "photoelectric effect."

The work for which Einstein is best remembered is his theory of relativity. It has been used to explain such things as gravitational lenses, time dilation, and the gravitational phenomena connected to black holes.

Einstein's Nobel Prize

In 1905, Albert Einstein was trying to explain how the photoelectric effect worked. This is a process whereby electrons are emitted from matter when it is bombarded with electromagnetic radiation (light). At that point in history, light was thought to travel only as a wave. However, if that were true, the more intense the light (or stronger the wave), the more energy the emitted electrons would have. Experiments showed that this was not true; in fact, the energy of the emitted electrons depended on the wavelengths of light being radiated. Einstein explained this phenomenon by pointing out that light acts as a wave, but it can also act as a particle, called a *photon* or a *quantum* of light. This double nature of light became known as *wave-particle duality*. Einstein explained that each photon has an energy level related to the wavelength of the light. Matter absorbs the quanta of light, and if there's enough energy, it then emits electrons. So, for example, when light strikes molecules of gas in space, those molecules absorb the energy of the light. If conditions are right, they emit electrons and the whole gas cloud glows.

The Life of a Genius

Albert Einstein was born in Germany in 1914 and from an early age showed interest in science. He studied mathematics and physics in college and then spent several years working as a patent officer after immigrating to Switzerland. His work involved evaluating products that used principles of electromagnetism, including those that transmitted signals. During this time he began working on his PhD in physics, and received his degree in 1905. He took a job as a lecturer at the University of Bern and taught physics for some years until he went to the Kaiser Wilhelm Institute for Physics.

Einstein spent much of his career working on his theories of relativity. As part of this work, he suggested that light from a distant object would become distorted as it passed near a gravitational field. This idea was put to the test during the solar eclipse of May 29, 1919. British astronomer Sir Arthur Eddington's photographs confirmed Einstein's hypothesis—light from distant stars appeared to be bent by the Sun's gravity. This was the first observed instance of gravitational lensing.

In 1933, with the rise of the Nazis to power, Einstein and his family immigrated to the United States, where he began doing research at Princeton University's Institute for Advanced Study.

He became a citizen of the United States in 1940 and spent the rest of his life studying relativity, quantum mechanics, principles of gravitation, and many other topics. He died in 1955, and his brain was preserved for research. Recently, neuroscientists searching for the origins of genius announced that Einstein's brain had several structural differences from ordinary brains. Whether or not those differences contributed to his genius remains to be proven.

Einstein's Theory of Relativity

There are actually two theories of relativity developed by Einstein. General relativity says that space and time make up spacetime. It can be affected by gravity, which curves space-time. It can also be affected by the presence of matter and energy and by the momentum of matter. This was a very new idea in the early twentieth century, and it required astronomers to rethink their perceptions of the universe. From our point of view here on Earth, space and time are locked and can't change. Yet Einstein suggested that, under the influence of gravity, space can expand, contract, or curve. Time, he said, depends on who is observing it and under what conditions observations are made. It seemed to many physicists like a chaotic way to run a cosmos. Yet, when you look at it more carefully, the ideas make perfect sense because a massive object curves both space and time. Depending on where you are in relation to that object, your sense of time can be affected. In other words, it's all relative.

Today, astronomers use the principle of relativity whenever they want to study objects that are moving through gravitational fields—such as light from a distant quasar being deflected by the gravitational pull of a galaxy cluster or a black hole.

Special Relativity

Science-fiction writers have a lot of fun with Einstein's theory of special relativity. It explains what happens when you have motion in two different frames of reference, particularly when one object is traveling at close to the speed of light. The laws of physics remain the same for each frame of reference. The speed of light is the same in both. If you were on a spaceship traveling at just under light speed, your local time would seem very normal to you. The clock would be ticking as it always does, and your motions on the ship would be completely ordinary. However, your friend back on Earth would see you speed away, and if he or she could see your clock, it would appear to be ticking faster than clocks on Earth. Conversely, you'd see Earth clocks moving more slowly, yet to your friend, his clocks would be telling time normally. If you raced out on a ten-year trip to space and back, you'd be ten years older, but your friend who stayed behind would be more than thirty years older! This effect is called *time dilation* and is one of the many implications of special relativity. Robert A. Heinlein's (1907–1988) novel *Time for the Stars* uses time dilation as a plot device to examine how people on a spaceship moving at near the speed of light age differently than their friends and family left back at home on Earth.

JOCELYN BELL BURNELL

Meet the Discoverer of Pulsars

In 1967, a young postgraduate student at the University of Cambridge in England was working with her advisor on a project to build a radio telescope to study quasi-stellar radio sources, commonly called quasars. Some of her data, though, looked odd. Strange bursts pulsated with a very regular rhythm. The student tracked these bursts across the sky until they disappeared. Day after day, she kept finding them in her data. Her advisor insisted that the signal was man-made or due to an anomaly in the equipment, but the student persisted in exploring it. Student and teacher checked out everything they could think of that might be causing this strange signal—ranging from automobile interference to signals from police cars to nearby radio and TV stations.

When a second pulsating source was discovered elsewhere in the sky, they had to take it seriously and considered the idea that it could be from a naturally occurring object or event in the universe. Jokingly they speculated that the bursts might be coming from intelligent life elsewhere in the universe, so they called the source LGM-1, which stood for "Little Green Men."

The student had, without knowing it, discovered the first pulsating radio source, later dubbed a pulsar. Her name was Jocelyn Bell (1943–), and her advisor was Antony Hewish (1924–). He later went on to win the Nobel Prize for his part as advisor of her work, and she was honored for her discovery with prizes throughout her career, although she did not share Hewish's Nobel Prize. Eventually she was named to the Royal Honors List, ultimately becoming Dame Commander of the Order of the British Empire.

A pulsar is a rapidly spinning neutron star that sends a beam of radiation out as it turns. If the beam sweeps across Earth's field of view, then its signal is detected as a quick, rapidly repeating blip.

The Science Behind LGM-1

The object that Jocelyn Bell Burnell discovered is a neutron star. These dense spheres of *degenerate matter* form when a massive star explodes as a supernova, or when a white dwarf star in a binary system accretes too much matter and collapses. What's left of the core of the star condenses to an object about the size of a city, spinning incredibly fast and emitting a powerful blast of radiation. If we happen to be in the line of sight of the pulsar's "beam," then our radio detectors pick it up as a very fast repeating signal. The first pulsar that Bell found is called PSR 1919+21, and its signal repeats precisely every 1.33 seconds.

From Ireland to the Cosmos

Dame Jocelyn Bell Burnell was born as Jocelyn Bell in Northern Ireland. Her father encouraged her to pursue academics, and she fell in love with astronomy at a young age. She went to the University of Glasgow in Scotland, earning a physics degree. At the age of twenty-three, she went to Cambridge to pursue her graduate work. There, she joined Antony Hewish and his team working on a large radio astronomy detector. That's when she started finding the mysterious signals on the many printouts of data the telescope produced. The media had a field day with the discovery of pulsars. Even though the discovery was Bell's and the original paper had her name on it as a co-discoverer, she was snubbed by the Nobel Prize committee for an award. The award to Hewish looked as if he was taking credit for her work, and that raised a firestorm of controversy over whether or not Bell was unfairly deprived of the honor of her discovery. However, she has consistently supported the Nobel Committee's decision and has stated that it was appropriate.

In later years, Bell continued to work in x-ray astronomy at Mullard Space Science Laboratory in England and pursued gamma-ray astronomy at University of Southampton. She also worked as a senior researcher at the Royal Observatory in Edinburgh, Scotland; taught at Open University; was a visiting professor at Princeton; and most recently took on a visiting professorship at Oxford. She continues to work to improve the status of women in the sciences.

Pulsar Research

Since the time of Bell and Hewish's work, astronomers have found hundreds of pulsars throughout the Milky Way Galaxy and in globular clusters. The big question is what happens to pulsars; the current answer is that they eventually slow down. Astronomers can detect that slowing as a lengthening of time between each of the pulses.

Pulsar Planets!

Jocelyn Bell Burnell's discovery opened up the strange world of the neutron star. The environment around one of these odd objects isn't the most hospitable in the universe. The neutron star itself has such a massive gravitational pull that if a planet got too close, it would break apart. In addition, the strong beam of radiation streaming from the neutron star would effectively fry any life that existed on such a planet—providing there was anything left after the planet weathered the supernova explosion that created the neutron star in the first place. Yet, in 1992, astronomer Aleksander Wolszczan (1946–) was the first to discover planets orbiting pulsars. There are currently six candidate objects to be studied to see if they are possibly pulsar planets. At least one of the pulsars has a circumstellar disk, likely made of heavy elements created when the parent star went supernova.

VERA COOPER RUBIN

Finding Dark Matter

In the 1970s, astronomer Vera Rubin (1928–) and her colleague Kent Ford (1931–) were trying to solve an interesting problem in astronomy. They wanted to figure out how mass is distributed throughout the galaxy. They started by looking at the motions of stars in the Andromeda Galaxy. The idea was that stars near the center of the galaxy should orbit faster than those on the outskirts. However, the observations showed something completely unexpected. When the team compared the stars' orbits at the heart of the galaxy with those farther out, they all seemed to be moving at about the same speed.

To confirm their findings, the team looked at other spiral galaxies, and found the same results. Something was causing all the galaxy's stars to rotate at close to the same rates, even though they were at different distances from the center of the galaxy. What was happening? Clearly something massive was affecting the orbital velocities of the stars. It was unseen but had enough gravitational pull to affect stellar motions. Rubin and her colleagues had found the first hints of what came to be known as *dark matter*.

From Childhood Dreams of Stars

Vera Cooper Rubin grew up with an interest in the stars. She built her own telescope with her father's help, and when it came time for college, she went to Vassar and studied astronomy. Once she got out, she attended graduate school at Cornell (after being turned down by Princeton because of her gender) and studied physics. She earned her PhD in 1954 and spent the next few years teaching at Georgetown and raising a family. Her main interest was in the dynamics of galaxies—the motions of their stars and other materials. That's what led her and her colleague Kent Ford to start studying distant galaxies.

They assumed that the core of a spiral galaxy has the most stars, which means it contains the most mass and thus exerts the most gravitational force in the system. Rubin and Ford started looking at the spectra of the stars in galaxies, using the Doppler Effect to determine their speeds. They calculated rotation curves of the motions of stars in the galaxy. That's when they discovered the odd match of velocities between the outer and inner stars.

Rotation Curves

Sometimes discoveries in astronomy come under unexciting names like "light curves" or "rotation curves." Light curves—which are simply graphs of the light streaming from distant objects—often reveal the most fascinating things about their objects. For example, the light curve of an asteroid can reveal whether the asteroid is tumbling in its orbit, whether it has light and dark areas on its surface, or even if it has a lumpy shape. The rotation curve of a galaxy, which is what Vera Rubin and Kent Ford studied, reveals the motions of stars in the galaxy and how they vary from place to place within the galaxy. To get that information, an astronomer takes spectra of the stars in a galaxy and then measures their Doppler shifts to determine their velocities. Put the spectra together from stars across a galaxy, and you get a graph called a rotation curve.

Rubin concluded that there was something massive and unseen in the galaxies they studied. She calculated that those galaxies had at least ten times as much of this unseen "stuff" as they did of the bright stars and nebulae. In other words, the bright material in a galaxy isn't the only thing there. She looked over research done by astronomer Fritz Zwicky (1898–1974), whose findings she had studied in graduate school. He was a Swiss astronomer who suggested in 1933 that supernovae could explode and that what was left would collapse into dense balls of neutrons (this theory was further solidified by Jocelyn Bell's discovery in 1967 of pulsars). He went even further and suggested that there could be mass in the universe that was largely unseen. Zwicky's studies of the Coma Cluster of galaxies found that there was much more mass among those collected galaxies than could be accounted for by the luminous stars and nebulae. If that unseen mass didn't exist, those galaxies should fly apart. The gravity of *something* was holding them together. He called this mysterious, invisible stuff dark matter, and this was what Vera Rubin remembered as she and her colleagues puzzled over the odd galaxy dynamics they had discovered.

Could it be that dark matter permeated all galaxies, and the universe itself? That's the question Rubin busied herself with, and in time, her work showed that Zwicky's dark matter does exist. For her part in the observations that confirmed the existence of dark matter (even though it still hasn't been directly detected), Vera Rubin has earned a number of prizes and honors. She continues to do research at the Department of Terrestrial Magnetism at the Carnegie Institution of Washington and focuses on the dynamics of what are called low surface brightness galaxies. These are dwarf galaxies that are diffuse and dim with few stars. They appear to have large amounts of dark matter.

Milky Way Star Motions and Dark Matter

The discovery that dark matter permeates the universe has astronomers eyeing our own galaxy more carefully. Exactly how massive is the Milky Way? It is estimated to have about a trillion solar masses of material we can directly detect. That's not the same as a trillion stars, however. It's the amount of all material in the galaxy expressed in units of solar mass. Furthermore, velocity measurements of stars in our galaxy using radio astronomy techniques seem to indicate that there is a huge mass of unseen material in addition to the stars, planets, and nebulae that we can see. Presumably this "stuff" is dark matter. Some astronomers have estimated that it could make up 95 percent of the mass of the Milky Way and could be equal to three trillion additional solar masses. Discovering the exact amount of dark matter in our galaxy is a subject of ongoing research for astronomers measuring the mass of the Milky Way.

CLYDE TOMBAUGH

From Kansas to Pluto

Astronomers used to be thought of as lone explorers sitting at the telescope night after night, just waiting to make the next big discovery. Nowadays, they work in teams, sometimes in multinational collaborations, each one contributing to the work of discovering and explaining the cosmos. In 1930, however, there *was* a lone explorer sitting at a special instrument at Lowell Observatory in Flagstaff, Arizona. His job was to take plates of the sky and then compare them to see if anything moved between exposures. His name was Clyde Tombaugh, and his painstaking work resulted in the discovery of Pluto, orbiting in the distant reaches of the solar system.

The Man Behind the Discovery

Clyde Tombaugh was born in Illinois in 1906, the son of a farmer and his wife. He and his father were both avid amateur observers, and Clyde often described himself as a young man building telescopes out of whatever he could find and grinding his own lenses. He wanted to attend college, but his family didn't have the money to send him, so he continued building telescopes and making observations of Mars and Jupiter. He sent some of his best ones to Lowell Observatory, hoping to get some advice from the observatory staffers. Much to his surprise, they contacted him with a short-term job offer. They needed an amateur astronomer to help operate their telescope. Tombaugh took the train to Flagstaff in mid-January 1928 and was met by astronomer Vesto Slipher (1875–1969). The job started right there, and Tombaugh ended up staying for thirteen years.

In 1928 he began doing planet searches. In particular, the director wanted him to find the mysterious Planet X that might exist out beyond Neptune. The project was spurred by Percival Lowell, who had founded the observatory but died in 1916. Lowell had been extremely interested in finding this unknown world and started a search program about a decade before he died.

Percival Lowell

If it hadn't been for Percival Lowell's fixation on discovering a new planet, Clyde Tombaugh might still have been well known as the discoverer of many other astronomical objects. However, Lowell's single-minded determination in funding such a planet search was the push needed to discover Pluto. Sadly, Lowell is sometimes better known as the man who speculated wildly about life on Mars. He founded Lowell Observatory for the express purpose of observing Mars and finding Martians. For more than twenty-three years before he died, Lowell and his colleagues used his observatory to further study the Red Planet, and to eventually search out the dwarf planet Pluto.

Once at Lowell Observatory, Tombaugh began using an astrograph to take photographs of sections of the sky where the hidden planet was thought to be. He'd take a set aimed at one place one night, and then a few nights later, he'd take another set of the same place. He would then compare them using what's called a blink comparator. This instrument allowed him to switch very quickly between one image and another to see if anything moved. If it did, Tombaugh made a note of it. It took him weeks to examine each plate that he took. Eventually, he found a very dim object that appeared to jump from frame to frame in a set of observations he'd made a week earlier. Further observations allowed an orbit to be calculated, which turned out to lie beyond Neptune. Tombaugh made his discovery on February 18, 1930. When the discovery was announced on March 13, it electrified the world. Pluto was the first planet to be discovered since Neptune had been found in 1846, and it was the first to be found by an American astronomer. The new planet was named Pluto, and it made Tombaugh famous.

During his search for the planet, Tombaugh also discovered more than 800 asteroids, hundreds of variable stars, and photographed such objects as star and galaxy clusters. In later years, Tombaugh went on to become a college teacher in Flagstaff and at the University of California at Los Angeles. He then worked at the White Sands Missile Range in the Ballistics Research Laboratory before taking a teaching position at New Mexico State University in 1955. There he spent time building up the astronomy department and its facilities before retiring to a life of stargazing and public lectures about his work. Clyde Tombaugh died in 1997, leaving behind a solid legacy as an observational astronomer. In his honor, some of his ashes are on their way to Pluto as part of the *New Horizons* mission to the outer solar system.

Lowell Observatory

This observatory set on a mesa called Mars Hill outside of Flagstaff, Arizona, is still very active these days, more than a century after it was built. Its telescopes still scan the heavens each night and members of the public are allowed to look through them. It has expanded out onto nearby Anderson Mesa and does work in partnership with the United States Naval Observatory and the Naval Research Laboratory. In addition, Lowell has sites in Australia and Chile and has opened up a new facility called the Discovery Channel Telescope about 65 kilometers south of Flagstaff.

MIKE BROWN

The Guy Who Helped Demote Pluto

There's a planetary scientist who cheerfully claims to have killed Pluto. His name is Mike Brown (1965–), member of a team that has discovered more outer solar system worlds than anyone else. In his defense, he claims that the distant little world deserved its fate. He even wrote a book about the subject, *How I Killed Pluto and Why It Had It Coming*, in which he explains the reasons why Pluto was demoted to dwarf planet status in 2006.

The Pluto-Killer Himself

Mike Brown is a professor of planetary science at the California Institute of Technology (CalTech). He was born and raised in Huntsville, Alabama, and went to college at Princeton and the University of California at Berkeley before moving to CalTech. His work in solar system exploration has netted him many awards, including *Time* magazine's 100 Most Influential People, but on his web page, he notes that the most important honor for him was winning an honorable mention in his fifth-grade science fair. He and his family were greatly amused when *Wired Online* voted him one of the Top Ten Sexiest Geeks in 2006.

Mike Brown and his team of world-hunters are painstakingly searching out tiny places in the outer solar system called Trans-Neptunian Objects (TNO). These lie out past the orbit of Neptune. They include:

- Quaoar
- Sedna
- Orcus
- Haumea
- Makemake
- Eris
- Pluto

Eris is actually larger than Pluto, and it's the world that started the whole "Pluto Is a Planet/No, It's Not" controversy that led to that world's redesignation as a dwarf planet.

Finding the Pluto-Killing World

Mike Brown, Chad Trujillo, and David Rabinowitz make up the TNO-hunting team that discovered Eris. They imaged it in 2003 during a sky survey as part of their systematic search for outer solar system bodies using the Samuel Oschin 1.2-meter telescope at the Palomar Observatory in southern California. Eris didn't get flagged at that time because it wasn't moving fast enough to be noticed as a potential TNO. Ultimately it turned up when the team analyzed their data. They did follow-up observations to calculate the object's orbit and figure its distance. When the team was sure of its discovery, they announced it on July 29, 2005, about the same time as they revealed two other worlds (Haumea and Makemake).

It took so long to find Eris because it moves very slowly across the backdrop of the sky. Data analysis and further observations revealed that not only is Eris larger than Pluto, but it also has its own moon, now known as Dysnomia. The world is named after the Greek goddess of discord and strife, which is perfectly appropriate, given the great controversy that arose as astronomers decided that Pluto was no longer a planet and would, along with Eris, henceforth be classified as a dwarf plant.

About Eris

What do we know about this world that helped kill off Pluto's planetary status? Eris appeared to be larger than Pluto when it was discovered, and people called it the "Tenth Planet" for a while. That nickname spurred discussion of just how many planets are out there, which then raised the question, "Well, what's a planet?" That's how the whole Pluto reclassification thing got started. Eris is also called a plutoid, which is another way to describe a TNO that's a dwarf planet. Its diameter, as measured by the *Hubble Space Telescope*, and by a ground-based observation from LaSilla in Chile, is about 2,300 kilometers, which makes it about the same size as Pluto. Like other objects in the Kuiper Belt, Eris is probably made of half ice and half rock inside, with a surface covered largely by nitrogen ice with a smaller amount of methane ice. That's a lot like Pluto!

Hunting for planets in the outer solar system is not an easy task. Clyde Tombaugh found that out in 1930 when he set about trying to find Planet X (later named Pluto). Objects in the outer solar system are dim and small, and because they orbit so far from the Sun, their orbits are very large. This means that they don't move very quickly. Tombaugh had to examine many photographic plates, comparing them to each other, before he could detect Pluto's plodding motion across the sky.

The same is true of today's modern TNO-hunters. They must take painstaking survey observations over many nights to catch a glimpse of a dim and distant object. The farther away the objects are, the harder it is to detect their motion. Moreover, the surfaces of these objects are not very bright, which makes them harder to spot. Luckily, these searches can now be automated. For example, the Samuel Oschin Telescope can operate in what's called a point-and-track mode, in which it locks onto a specific area of the sky and takes an exposure. Then it looks at other areas of the sky and takes a pre-set number of images before returning to the original area. If anything has moved in any of the views, it gets tagged for further review. This works really well in the search for dim, distant solar system objects such as comets and asteroids, as well as TNOs and objects out in the Kuiper Belt region beyond Neptune.

More New Worlds

Haumea and Makemake, the two worlds announced at about the same time as Eris, are also orbiting the Sun out beyond Neptune.

1. Haumea is an ice-covered rocky body and is named after the goddess of the Big Island of Hawaii. It has two moons called Hi'aka and Namaka. This little world is oblong-shaped, somewhere around 2,500 kilometers long and about 1,500 kilometers across. It lies at a distance of 35 AU, and its orbit is 283 Earth years long.

2. Makemake is an ice-covered dwarf planet with dimensions of about 1,500 kilometers by around 1,400 kilometers. At the most distant point of its orbit, it lies 53 AU from the Sun and takes nearly 310 Earth years to make one orbit.

ASTROPHYSICS AND ASTRONOMY

The Physics of the Universe

Astronomy is, at its heart, the scientific study of objects and events in the universe. It's a rich science, divided into two major branches:

1. *Observational astronomy*, which is concerned with gathering as much information about objects in the universe as possible. Think of it as the data-gathering part of astronomy.
2. *Astrophysics*, which applies physics to explain the properties, interactions, and evolution of planets, stars, the interstellar medium, nebulae, galaxies, and other objects in the distant reaches of the cosmos. Astrophysicists also apply aspects of chemistry, electromagnetism, particle physics, and other disciplines to explore and explain objects and processes in the universe.

Astronomers (who are also usually astrophysicists) make their observations using observatories outfitted with instruments sensitive to light from different parts of the electromagnetic spectrum. Using the data from their observations, these scientists come up with explanations for what's happening in the universe. *Theoretical astrophysics* uses models, statistics, and simulations to explain objects in the universe and predict what they might do in the future.

Seeing the Light

Light is one of the most fundamental parts of the universe and is the standard "currency" in astrophysical research. Astronomers study light emitted and reflected from objects in order to understand more about them and their environments. Light can act as a particle, called a photon, or it can travel through space as a wave. This dual nature of light is central to how we detect objects in the universe. We can collect photons using cameras, but we can also measure wavelengths of light.

The word *light* is usually used to describe light our eyes can see. We evolved to be most sensitive to visible emissions from the Sun. But those are only a small part of the *electromagnetic spectrum*—the range of all possible light given off, absorbed, and reflected by objects in the universe. Most of the rest of the spectrum is invisible to us because it's in the form of x-ray, ultraviolet, radio, infrared, and microwave emissions. Astronomers use specially sensitive instruments to detect them.

Infrared Astronomy

For centuries, astronomy was a visible-light science. In the 1800s, scientists began measuring and analyzing other wavelengths of light, starting with infrared, also known as thermal (heat) radiation. Anything that is even slightly heated gives off infrared radiation (often referred to as IR). Unaided, we may not be able to see what is doing the heating, but IR detectors can "lift the veil" for us.

A good example is a cloud of gas and dust surrounding a newborn star. IR-sensitive detectors zero in and show us the star. They let us see the region around a black hole or peer into the depths of a cloud hiding a star that's about to die. Much infrared astronomy is best done from space, since Earth's atmosphere absorbs a lot of incoming thermal radiation.

Infrared Telescopes

The most famous infrared missions to date are the *Spitzer Space Telescope* and the European Space Agency's *Herschel Space Observatory*. There are some high-altitude observatories such as Gemini (in Hawaii and Chile) and the European Southern Observatory (also in Chile) that do good infrared observing because they're located well above much of Earth's atmosphere.

Ultraviolet Astronomy

Ultraviolet astronomy focuses on light that is more energetic than infrared or visible light. Ultraviolet (UV) is also absorbed by Earth's atmosphere, so the best observations are done from space. What gives off UV in space? Hot and energetic objects do. This includes young stars and superheated interstellar gases. The Sun gives off UV, which is what burns your skin if you stay outdoors without good sunblock.

UV Telescopes in Space

The best-known UV detectors are the *International Ultraviolet Explorer* mission, the UV-sensitive instruments first launched aboard the *Hubble Space Telescope*, and the *Galaxy Evolution Explorer* (GALEX).

Radio and Microwave Astronomy

Early in the twentieth century, an engineer at Bell Labs named Karl Jansky (1905–1950) pointed a radio receiver at the sky and inadvertently became the first person to discover naturally occurring radio signals from an object in space. The emissions he found came from the center of the Milky Way Galaxy. Today, radio astronomers use vast arrays of radio dishes and antennas to detect signals from a wide variety of objects. These include superheated shells of plasma (energized gases) emanating from the cores of galaxies, shells of material from supernova explosions, and the microwave emissions from the vibration of interstellar molecules in clouds of gas and dust in interstellar space or in planetary atmospheres. In addition, atmospheric researchers use radars and radio dishes to study interactions of Earth's upper ionosphere with the solar wind and radars to map such places as the cloud-covered surface of Venus and Saturn's moon Titan.

The Father of Radio Astronomy

Karl Jansky was born and raised in Oklahoma, where his father taught engineering and encouraged his children to dabble with radio sets. The science of radio waves intrigued Karl, and he studied physics in Wisconsin. He was hired by Bell Labs to do research on properties in the atmosphere that affect radio transmissions. It was during this time that he built an antenna that his fellow workers nicknamed "Jansky's Merry-Go-Round"—a rotating contraption that allowed him to seek out signals from naturally occurring processes and objects.

Jansky's discovery of the Milky Way's emissions clearly pointed the way toward a new branch of science. He should have continued his research, but his employer wanted him to work on other projects, so that was the last astronomy work he ever did. That didn't stop others from taking up radio astronomy, and eventually it emerged as a subdiscipline of astronomy.

X-Ray and Gamma-Ray Astronomy

The most energetic objects, events, and processes in the universe give off x-rays and gamma rays. These include supernova explosions such as the one that created Cygnus X-1, the first x-ray source to be discovered; high-speed jets of matter streaming from the cores of active galaxies; and distant powerful explosions that send both x-rays and gamma-ray pulses across space. To detect and study these powerful emitters, astronomers have used space-based observatories such as the *Chandra X-Ray Observatory*, the *Roentgen Satellite (ROSAT)*, and *XMM-Newton* for x-rays, and the *Compton Gamma-Ray Observatory (CGRO)*, the current *Fermi* mission, and the *Swift* satellite for detecting gamma-ray emissions in the universe.

Did You Know?

NASA's *Fermi* satellite not only studies strong gamma-ray bursts from the distant reaches of space; it also captures gamma-ray flashes that occur right here on Earth near the tops of violent thunderstorms.

One of the most startling discoveries in microwave astronomy came in 1964 when scientists Arno Penzias (1933–) and Robert W. Wilson (1936–) detected a glow of background radiation at microwave frequencies. It seemed to come from everything in the universe. This so-called "cosmic background radiation," or CMBR, is the leftover glow from one of the earliest epochs of cosmic history shortly after the Big Bang occurred.

Spectroscopy in Astronomy

Astronomers pass light from an object through a specialized instrument called a spectroscope. Think of it as a very special type of prism, but instead of creating just the array of colors we see, it separates light into very fine divisions called a spectrum that extends far beyond what our eyes can detect. The chemical elements in an object emit or absorb wavelengths of light, which shows up in a spectrum as a glowing bar of light or a dark "dropout" line. Therefore, you can imagine spectra as cosmic bar codes; encoded in those bars is information about the chemical composition, density, mass, temperature, velocity, and other characteristics of planets, stars, nebulae, and galaxies.

Mining the Universe

Astronomical instruments are producing prodigious amounts of data each day and night. Some information gets analyzed right away, but there are huge databases of observational information just waiting to be studied, particularly from the many automated sky surveys that sweep the sky on data-gathering expeditions. This has opened up new avenues of research through data-mining. Trained experts go through these vast treasure troves of information and find new data points to study and correlate with other observations. For example, the *Hubble Space Telescope* has taken many images of distant galaxies. Once the astronomers who proposed the observations are through with the data, they release it to archives, where anyone can study it. Other astronomers who study galaxy shapes and evolution saw those images and data as a gold mine of information. Classifying those galaxies by their shapes is a necessary part of the research into galaxy evolution. So, astronomers have devised machine-enabled searches to pick galaxies out in the images and sort them by shape. There's also a Citizen Science project called Galaxy Zoo that allows members of the public to examine these HST images and classify the shapes of the galaxies in them. The work they do helps astronomers understand the origin and development of galactic objects in the early universe.

ASTROBIOLOGY

The Origins and Evolution of Life in the Cosmos

From the moment people first realized that there were planets around other stars, we have wondered if those places contain life. The civilizations of antiquity had stories about alien worlds populated with gods, but these were more mythological than science-based. In fact, the Greek philosopher Aristotle flatly ruled out the idea of life on other worlds, and his Earth-centered cosmology held sway for more than 1,200 years.

With the advances in scientific thought made possible by the Copernican Revolution, people began to understand that there could be other worlds "out there" and that the idea of life elsewhere in the cosmos is not so strange. The advent of the Space Age allowed humans to put space probes in orbit around other worlds and eventually to land on them. The question of whether or not life exists elsewhere fostered a new discipline involving chemistry, physics, astronomy, molecular biology, planetary science, geology, and geography. This branch of science is called astrobiology.

The Roots of Life

Astrobiology is the study of the origin, evolution, distribution, and future of life in the universe. This scientific discipline uses research tools to search out life-friendly neighborhoods on the other worlds of our solar system. Since the discovery of the first exoplanets (worlds orbiting other stars) in 1995, astrobiologists have been devising ways to search for signs of life on those worlds, too.

Life on Earth has a long history, rooted in a complex biochemical evolution that started with simple compounds and molecules that found their way to our planet's oceans and surfaces. Under the right conditions, those compounds combined and eventually gave birth to the first primitive forms of life. The prebiotic chemistry known to exist on early Earth illustrates what the environment was like during the planet's formative period. Furthermore, astrobiologists can apply what they have learned about our own planet to analyze conditions on Mars or Titan, for example, to see if those worlds provide environments conducive to life.

Under the Sea

The deep sea is one of the least-explored places on our planet, and there are species of life deep in our oceans that have only recently been catalogued. How they live in such alien conditions could give us clues as to how life might exist in oceans on other worlds.

Twenty-First-Century Astrobiology Missions

Today astrobiology is a very active branch of scientific inquiry. In the United States, NASA funds the Astrobiology Institute, which was developed in the late 1990s to provide guidelines in the search for life elsewhere. A number of universities are also involved in the multidisciplinary science, and the European Space Agency is actively researching topics in astrobiology as well.

The Goldilocks Zone

Now that astronomers are finding planets around other stars, the search for life on those worlds brings new challenges to astrobiologists. One thing they want to know is whether or not a planet orbits in its star's habitable zone, or "Goldilocks Zone." This refers to the children's story where Goldilocks tries to find porridge that's just right: not too hot, not too cold. In this case, the Goldilocks Zone refers to a region around a star where a planet would be in just the right orbit for conditions to be warm enough to allow liquid water to exist on its surface and thus be a possible breeding ground for life. The next steps are to study the planet's size, its atmosphere, and other factors to see if life could be present. There may be such zones around stars where chemical compounds other than water could exist in liquid form, and it's possible that life forms could eke out an existence in those environments, too.

A large part of astrobiology is the study of what are called *extremophiles*—life forms that can exist and thrive in really hostile conditions. For example, there are microbes that can live quite happily in hot springs or in the deep sea in proximity to underwater volcanic eruptions. Other tiny life forms adapt well to freezing conditions or in places where water is extremely scarce. One of the more interesting life forms is a species of deep-sea worm that appears to live quite nicely while burrowed into methane ice deposits. The existence of these and other extremophiles gives some hope that life could be flourishing elsewhere in the solar system under similar conditions.

As astronomers search for life on worlds beyond our solar system, astrobiologists are focusing on Sun-like stars because they could well have evolved similarly to our own. The main assumption is that any life out there is carbon-based (as we are). It's not so far-fetched—carbon is a very versatile element, and it combines with many other elements. It's also very common in regions where stars form and thus could be one of the main building blocks of life on other worlds. It's only a matter of time before we find a Sun-like star and its orbiting world with traces of water in its atmosphere.

The Search for Life on Mars

Right now, Earth is the only place where life is proven to exist. However, scientists have long cast their gaze on Mars, wondering if life could have arisen there, too. To find out, they've sent dozens of probes to the Red Planet. The most successful were orbiters that mapped the surface, joined by landers and rovers that settled on the surface to stir up the dust, study the rocks and ice, and sniff the atmosphere. The *Mars Curiosity Rover* and the *Mars Exploration Rover* are the two most recent missions to land on the planet. They're doing active geological studies aimed at finding out the history of water on Mars and looking for any chemical or physical signs that life could have arisen or still exists on the planet.

PLANETARY SCIENCE

Learning about Planets

Want to know how a planet formed? What shaped its surface? Where its moons came from? What will happen to it? Then you want to study planetary science. This field covers all objects in the solar system and how they came to be the way they are. What scientists learn in that field can be applied when they observe planets orbiting other stars, too. Planetary science began as a subdiscipline of astronomy but quickly branched out. Today, a good planetary scientist must know:

- Geology
- Geochemistry
- Geophysics
- Atmospheric science
- Glaciology
- Oceanography

If you're planning to go into this field, you'll even study a little spacecraft engineering in order to design and build instruments for planetary exploration.

Processes That Shape Planets

The basics of planetary science are similar to the principles of geology. Assuming that you're studying a world with a hard surface—a rocky planet or moon, an asteroid, or a place with an icy surface such as a comet or a frozen moon—there are several processes that can affect that place. The first is *tectonism*. This affects the outer layer of a planet or a moon, and it happens very slowly. It's caused by heat escaping from under the crust of the planet. As it does, it warps the surface, causing it to fold or create faults (fractures).

Worlds with Tectonism

On Earth, tectonism is responsible for mountain building, volcanism, and earthquakes. There is likely some form of tectonism at work on Mars and Venus, as well as on some of the icy worlds of the outer solar system, including Europa and Ganymede at Jupiter; Titan and Enceladus at Saturn; Ariel, Titania, and Miranda at Uranus; and Triton at Neptune. The *New Horizons* spacecraft will look for evidence of tectonics at Pluto when it flies by in 2015.

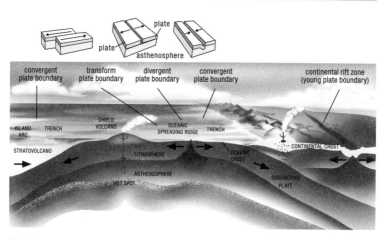

Earth's tectonic actions push plates together, fracture the surface with earthquakes, and force mountain-building activities that take millions of years to complete.

The second process that affects a hard-surfaced world is *impact cratering*. Debris has smashed into the surfaces of most solar system objects, ranging from Mercury to Mars, out to the moons of the gas giant planets. It's likely the dwarf planet Pluto and its cohorts in the Kuiper Belt also have impact craters.

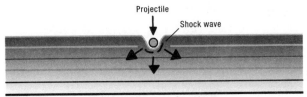

Contact/compression stage

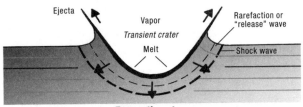

Excavation stage

End excavation stage/start modification stage

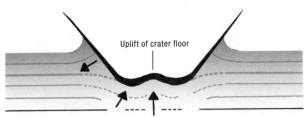

Modification stage

When a projectile hits a hard surface, it excavates a crater. The material it plows into reacts. Some of it is ejected away from the crater. If the impact is strong enough, the surface will rebound to form a small peak in the center of the crater.

ASTRONOMY 101

Much of the cratering occurred early in the solar system's history, during a period called the Late Heavy Bombardment. However, impacts continue to occur as errant solar system stuff crashes into planets, moons, and asteroids. The study of impacts gives us a handle on the collisional history of the solar system, particularly in the early epochs when smaller bits of solar system material merged to form larger worlds.

Impact on Jupiter!

Gas giants can experience impacts, too. In 1994, twenty-one pieces of Comet Shoemaker-Levy 9 plowed into Jupiter's upper cloud decks. The kinetic energy and heat from the collisions destroyed the comet pieces, leaving behind plumes of dark material that eventually faded. Since then, astronomers have seen other objects collide with Jupiter. It's very likely that because of its immense gravitational field, Jupiter acts as a sort of vacuum cleaner, sweeping up or deflecting incoming asteroids and comets that might otherwise pose a threat to planets in the inner solar system.

A third process that affects planets is *weathering*. It occurs when surfaces are affected by interactions with water, wind, and vegetation. We see plenty of all three on Earth. Flowing water and wind erode rock, and plants such as mosses can break down rocks as well. Chemical weathering is also at work across our planet. It occurs when acid rain dissolves rocks, for example. Windstorms scour the landscapes, and the freezing-thawing cycles that accompany winter and summer weather can crumble rocks, too. There are also signs of weathering on the surfaces of Mars and Venus, caused by contact with the atmosphere, and on Mars by the action of liquids that flowed across its surface in the very distant past.

Space Weathering

The surfaces of worlds exposed to space also undergo a type of weathering. This includes the Moon, Mercury, asteroids, comets, and many moons in the outer solar system. On the Moon, micrometeorites plow into the surface, melting and vaporizing the soil and churning the dusty regolith. On icy bodies, weathering can take place when ices in their surfaces are bombarded by solar ultraviolet radiation or high-energy particles such as cosmic rays. This causes the ices to darken and explains why many ice-covered moons in the outer solar system appear so dark.

Volcanism is another major force that changes the surfaces of solar system worlds. We are most familiar with the volcanoes on Earth, which belch lava and clouds of poisonous gases. There are volcanoes on the continents as well as in the deep ocean. Essentially they convey heat from beneath the crust of our planet and in the process resurface the land and seascapes.

Volcanoes in Space

Volcanism has played a role in shaping Mercury, Venus, Mars, and many of the icy worlds in the outer solar system. On the inner planets, the volcanism is basaltic—meaning that volcanoes spread molten rock across the landscapes. On the outer moons, frozen and slushy ice (and sometimes very cold water) is spewed onto the surface. As with volcanoes elsewhere, heat from within the world is melting subsurface ices and forcing them to erupt through what are called *cryovolcanoes*.

Planetary Atmospheres

The study of atmospheric blankets around planets (and some moons) focuses on the processes that create and sustain atmospheres, as well as their structures and effects on the planet. Atmospheric scientists also study interactions between a planet's blanket of air and its magnetic field. These days climate change is in the news, and this makes the study of Earth's long-term atmospheric changes and their effects on our climate all the more important. Various space agencies have launched satellites to monitor our warming atmosphere from space. Ground-based research also includes studying interactions between the global oceans and the atmosphere.

COSMIC TIME MACHINES

The World's Observatories

Modern astronomers use a wide variety of observatories to explore the universe. They range from research-grade university telescopes, eight-meter-wide professional instruments on remote mountaintops, and multi-dish radio arrays to space-based observatories and planet probes. There are more than 11,000 professional astronomers in the world today studying objects as close as the Moon and as far away as the first ripples of radiation detectable after the Big Bang. In addition, there are (by some estimates) more than half a million amateur observers watching the sky with binoculars, small backyard-type instruments, and even some amateur radio astronomy dishes. No matter what their size or where they're located, these cosmic time machines extend our vision across time, space, and the electromagnetic spectrum.

History of the Telescope

Our first astronomical instruments were our eyes. From the earliest sky-gazing activities in antiquity through the invention of the telescope in the 1600s, people simply had to observe objects in the universe without magnification. In 1608, the first instruments that could magnify a person's view were created. The invention of the telescope is often credited to Dutch opticians Hans Lippershey, Zacharias Janssen, and Jacob Metius. These telescopes were likely originally used as spyglasses aboard ships or by generals during a battle, but it wasn't long before someone thought to look at the sky with these newfangled inventions. The news of these instruments

ASTRONOMY 101

reached astronomer Galileo Galilei, who promptly built his own instrument in 1609.

The first telescope was simply a lens through which light would pass into an eyepiece. Later on, astronomers began to build telescopes with metal mirrors at one end, inspired by a design by Isaac Newton. Newton's design was so useful that it is still used today in telescopes called Newtonian reflectors. The primary purpose of all these optical telescopes is and always has been to gather as much light from dim, distant objects as possible.

Over the next few centuries, people worked to improve the telescope, replacing metal mirrors with larger (and heavier) ground and polished glass ones. With the advent of electricity and automation, people began using motors to guide their telescopes through long observations. Today's modern optical (visible-light) telescopes are computer-guided creations that can be steered quickly to view an object or event occurring in the sky. Among the better known are:

- Mauna Kea in Hawaii, including the twin Keck telescopes, the Subaru Telescope, and the Gemini North telescope
- The European Southern Observatory in Chile
- The Australian Astronomical Observatory in Coonabarabran
- Palomar Mountain Observatory, Mount Wilson, and Lick Observatory in California
- The Kitt Peak National Observatory in Arizona

In addition, radio astronomers use dishes and antennas to detect the radio emissions reaching Earth from distant objects. Most, but not all, radio telescopes are arranged in arrays such as the Very Large Array (VLA) in Socorro, New Mexico, and the Atacama Large

Millimeter Array (ALMA) in Chile. Radio arrays are also being built in Australia and South Africa, including the Square Kilometer Array (SKA) and the Murchison Wide-Field Array (MWA).

Multi-Wavelength Instruments

There are hundreds of observatories spread across every continent on Earth and orbiting our planet. Most ground-based facilities are limited to detecting visible light or radio emissions coming from objects in the sky, while space-based instruments cover a wider range of emissions. Since the late nineteenth century, astronomers have been attaching instruments to ground-based telescopes, including cameras (which record light) and spectrographs (which break incoming light into its component wavelengths). In the past few decades, some facilities here on Earth have been designed and optimized for infrared observations. Infrared (IR) is absorbed by our atmosphere, so these observatories are generally located at high altitude in very dry climates. This allows them to detect near- and mid-infrared light. The twin Gemini telescopes at Mauna Kea in Hawaii and Cerro Pachón in Chile, as well as the Very Large Telescope in Chile, are good examples of such IR-enabled observatories.

Space-Based Observatories

Earth's atmosphere is the bane of astronomical observing. Motions in our atmosphere cause stars to appear to twinkle and images of planets to waver. Our blanket of air, which protects us by absorbing infrared, ultraviolet, x-rays, and gamma rays, also makes it very difficult to spot cosmic sources of these emissions. On top of that, all astronomers often have to deal with clouds, which block their view, and light pollution, which washes out the view of dim and distant objects. Some of these problems can be mitigated by locating observatories at high altitude and away from light sources, but an even better solution has been around for decades. Beginning in the 1960s, astronomers began placing observatories in space, which gives constant access to distant objects not always visible from Earth, or radiating in wavelengths of light not always detectable from the ground.

Early Astronomers

The oldest observatories on Earth had no telescopes. These ancient places were as much cultural icons as they were sky observing posts. They include such sites as:

- Stonehenge in Great Britain
- El Caracol in Mexico
- Angkor Wat in Cambodia
- Ujjain in India

Ancient astronomers used these places not just for collecting observations of the stars and planets but also for such practical purposes as determining calendrical dates important in both religious and civil events. The first places for scientific observations of the sky began in antiquity and were scattered throughout ancient Greece, the Middle East, and China. In the early Middle Ages, Moorish invaders built observatories throughout North Africa and Spain. Some of them still stand today. European observers didn't start creating observatories until Tycho Brahe built his observation post at Herrevad Abbey in 1570 in what is now Sweden.

Flying Observatories

Balloons and high-flying aircraft also act as observatories. The first balloon-borne astronomy facilities were lofted to high altitude beginning in 1957. Astronomy instruments have been carried up into the atmosphere, giving astronomers x-ray, microwave, gamma-ray, and infrared access to the cosmos. Flying observatories include the Kuiper Airborne Observatory (KAO), which was used until it was decommissioned in 1995, and the Stratospheric Observatory for Infrared Astronomy (SOFIA), which flies in a modified Boeing 747 and began its mission in 2010.

The Higher the Better

The world's highest observatories are on Mauna Kea on the Big Island of Hawaii at 4,205 meters (13,796 feet) and in Chile's Atacama Desert at 5,640 meters. This is where the Atacama Large Millimeter Array (ALMA) studies the universe in radio wavelengths. It has been joined by the Tokyo Atacama Observatory, a visible light/infrared-enabled observatory, which had its "first-light" observations in 2009.

HUBBLE SPACE TELESCOPE

From Techno-Turkey to Astronomical Icon

In April 1990, the *Hubble Space Telescope* (*HST*) roared into space, strapped into the cargo bay of the space shuttle *Discovery*. This joint program between NASA and the European Space Agency (ESA) was the culmination of decades of planning by scientists and engineers. It wasn't the first telescope lofted into space, but it was the first one designed to be serviced in orbit. As of 2012, *HST* has made well over a million observations, looking at such celestial targets as the Moon and planets to distant stars and galaxies in the universe. It was designed to be a multi-wavelength observatory, and through the years it has hosted instruments sensitive to optical, infrared, and ultraviolet light. Its contributions to astronomy are as far-reaching as they are spectacular, and there's enough data to keep student researchers and their mentors busy for years to come.

Hubble's Heritage

The *Hubble Space Telescope* is named for the astronomer Edwin C. Hubble (1889–1953), whose observations of distant galaxies helped expand our view of the universe. The idea for *HST* came from a German rocket engineer named Hermann Oberth (1894–1989), long considered the father of modern rocketry. He published a book in 1923 called *Die Rakete zu den Planetraümen* (*The Rocket Into Planetary Space*), which devotes a great deal of time to the idea of a space telescope shot into orbit by a rocket. His observatory was to be populated by human crews who would function as telescope operators for ground-based astronomers. The third father of *HST* was astronomer Lyman Spitzer Jr. (1914–1997). In 1946, Dr. Spitzer wrote a paper in which he described the great advantages of locating an observatory out in space beyond our turbulent atmosphere. It took many years before his idea was taken seriously, but finally in 1965, Spitzer was allowed to go ahead with scientific planning for what eventually became the *Hubble Space Telescope*.

Earth, We Have a Problem

Soon after its launch and deployment, the first images started coming back from the giant telescope. They looked weird and unfocused. After a great deal of discussion and experimentation, engineers determined that *HST*'s mirror had been ground incorrectly. It had an optical problem called *spherical aberration*. The news was devastating, and the telescope became a lightning rod for criticism, particularly from politicians. Maryland Senator Barbara Mikulski, who is now one of the observatory's biggest supporters, was one of the most vocal critics at the time, calling the telescope a "techno-turkey." Behind the scenes, engineers and optics experts worked to come up with solutions. Computer programmers devised ways to "deconvolve" the spherical aberration through complex mathematical means. For the first three years in orbit, *HST* was able to deliver scientific data but not the clear, sharp images users were expecting.

In December 1993, the first servicing mission to *HST* installed corrective optics—basically, it gave the telescope a badly needed pair of glasses. To install the corrective optics, called COSTAR, astronauts had to remove one instrument to make room. They also replaced the main camera with a new one that had its own updated optics. Those repairs brought the telescope back to "spec." The first images from the newly serviced telescope proved that *HST* could do the job it was built to do.

There were four more servicing missions to *HST* over the years, replacing instruments and other housekeeping items. The telescope is now good to go for at least another decade, until its successor—the infrared-optimized *James Webb Space Telescope*—is launched, perhaps in 2018. Eventually, *HST*'s systems will age and deteriorate, although this isn't expected to happen soon. When it does fail, NASA has plans to bring it back to Earth in a controlled re-entry that will safely avoid populated areas.

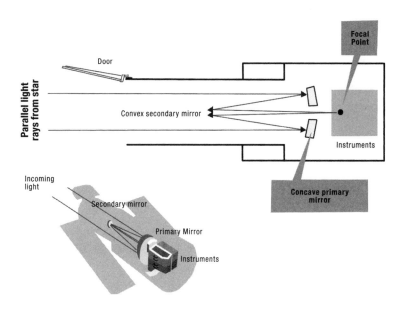

This diagram shows the main parts of the *Hubble Space Telescope*.

Astronomy's Key to the Universe

Over the years, the *Hubble Space Telescope* has tracked changes in the atmospheres of Mars, Jupiter, Saturn, Uranus, and Neptune. It has discovered more moons around distant Pluto. Beyond the solar system, *HST* has:

- Peered into the hearts of star birth regions
- Spied out a planet orbiting a nearby star
- Looked at the remains of Sun-like stars called planetary nebulae
- Watched as blast waves from a supernova crashed through clouds of interstellar gas and dust
- Measured accurate distances to faraway galaxies
- Searched out evidence for black holes at the hearts of galaxies
- Explored the earliest stars
- Helped refine the age of the universe

The telescope that was once a techno-turkey is now one of the most productive observatories in history, and it continues to be oversubscribed by astronomers wanting to use it. Its technological advances (particularly for the instruments needed to overcome its mirror problem) have spurred advances in ground-based observatories as well.

Bang for the Buck

The total cost of the *Hubble Space Telescope* from its starting date has been estimated at about $10 billion over thirty years. One scientist calculated that this would cost about two cents per United States taxpayer per week.

Controlling and Programming *Hubble*

The *Hubble Space Telescope* is controlled by engineers at the NASA Goddard Space Flight Center, while the Space Telescope Science Institute (STScI) is in charge of all science operations. STScI routinely takes applications for observing time from observers around the world, juggles them into a complex schedule, and then archives the data for the scientists as it is received. The astronomers then have exclusive access to their data for up to a year, which allows them time to analyze it properly. After that, the data are available for anyone to study. The institute's public affairs office regularly works with scientists to select images and stories to share with the public, and it maintains a website where anyone can enjoy and download *HST* images for personal use.

Hubble Trivia

The *Hubble Space Telescope* has transmitted well over 60 terabytes of data back to Earth. The telescope's users have published nearly 10,000 papers based on their observations, and there is plenty of data yet to be analyzed, with more to come.

THE *KEPLER* MISSION

Searching for Exoplanets

When you look up at the sky at night and see the stars twinkling, it's pretty easy to wonder if there are other planets like ours out there. Would they look just like Earth? Would they have life? What would such life be like?

The first step in answering those questions is to search out stars with planets. It's easy to do in our own solar system. We know where the planets are and are using every means we have to determine if any of the worlds circling the Sun (or other planets) have life. But when it comes to looking for worlds orbiting other stars, there's a big obstacle to finding them. Planets are very small and dim, and the light of their star hides them. As a result, astronomers have to use special techniques to find them.

The first confirmed detection of a world outside our solar system came in 1992 when astronomers found a planet orbiting a pulsar. In 1995, the first exoplanet orbiting a main-sequence star was discovered. Since then more than 900 exoplanets have been confirmed by both ground-based and space-based observations. In early 2013, astronomers announced that one in six stars in our galaxy likely has at least one planet associated with it and that many of those stars are not Sun-like but could be as exotic as white dwarfs—the remnants of stars that have entered into their extreme old age.

In the past few years, *Kepler* has been responsible for many planet discoveries, and is looking for planets in *Habitable Zones*, regions where water could exist as a liquid on the surfaces of terrestrial-type planets. Since life arose on our own terrestrial planet in the habitable zone around the Sun, it makes sense to look for planets in similar zones around other stars. The sizes of the planets already discovered and their compositions are revealing many similarities to our own solar system—and many differences. The study of those distant stellar systems, coupled with the exploration of our own piece of the galactic neighborhood, is providing many new insights into the general picture of star and planet formation (a topic explored more deeply in the Star Birth chapter).

Introducing the *Kepler* Mission

On March 6, 2009, NASA launched *Kepler* into an Earth-trailing orbit. Its assignment is to survey a small portion of the Milky Way, looking for stars with planets. To do this, the spacecraft is equipped with a 0.95-meter- (~3-foot-) diameter telescope that serves as a photometer. Think of it as a giant light meter that's sensitive to the dimmest flickers of light from stars and planets. Its perch in space removes the distortion that our atmosphere gives to incoming light, allowing the telescope to stare at distant stars without interruption. It concentrates on a patch of stars in a 3,000 light-year–wide region in the Cygnus-Lyra area of the sky. So far it has found more than 2,700 planet candidates, and more than a hundred of its discoveries have been confirmed as planets. It has also discovered brown dwarfs—objects too hot to be planets and too cool to be stars. *Kepler* mission data is archived for use by scientists doing follow-up observations in their own planet searches. The *Kepler* website at *www.kepler.nasa.gov* has instructions for obtaining the data.

Kepler's Target Zone

Kepler mission scientists chose stars that lie about 50 to 3,000 light-years away in the the Cygnus-Lyra region of the Milky Way to search for planets. They had several reasons for selecting that region: it's far enough away from the Sun so that it doesn't interfere with the telescope's instruments, the region is visible from the northern hemisphere where Kepler's partner ground-based telescopes are, it doesn't contain a lot of star-forming regions and giant molecular clouds, and the star density is high since it's close to the galactic plane. It's an ideal place to hunt for planets.

How does *Kepler* search for planets? It uses a technique called "the transit method." Here's how it works. When a planet crosses—or transits—between us and its star, it affects the star's brightness. Usually this is a very tiny amount of dimming, which is why the mission is equipped with an ultra-sensitive photometer that can detect these little flickers. If the dimming repeats itself over and over again in a regular way, then it is likely caused by a planet orbiting the target star. The object's orbital size is calculated from its period, and by using measurements of the planet's orbit and the properties of its star, astronomers can figure out if the newly discovered world is in the star's habitable zone, also called the Goldilocks zone.

The planet's orbit must be lined up edgewise between us and its star for us to be able to see it. Of course, not all planets are so fortuitously lined up, so the mission has to look at many hundreds of stars to find a candidate that has a planet (or planets) lined up precisely for detection. To find a planet that is an analog of Earth, that orbit has to lie in or very close to its star's habitable zone. Among other characteristics, the size and location of a star's habitable zone depends on its own temperature and state of evolution. A hot, bright star might have a habitable zone farther out than a dimmer, cooler one. So each star has to have its temperature and brightness measured, which gives a clue about the age of the star and its evolutionary history. As a result, in addition to uncovering the existence of planets, the *Kepler* mission is also aiding the understanding of how distant planetary systems form.

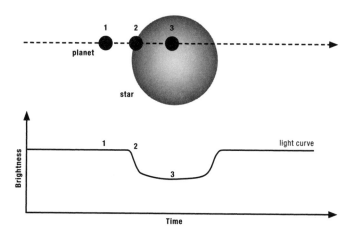

As a planet crosses between us and its star, it dims the star slightly. *Kepler* is sensitive enough to pick up that dimming in the light curve.

CoROT

Kepler has not been the only space-based mission looking for planets. The French Space Agency launched its *Convection, Rotation, and Planetary Transits (CoROT)* satellite in 2006. During its mission it found hundreds of objects that astronomers are working to confirm as planets. The process for *CoROT* planets as well as *Kepler* discoveries also involves repeated observations of planet candidates using ground-based observatories.

Kepler's Goals

The *Kepler* mission has the following scientific objectives to meet:

1. Primarily to determine the fraction of stars that have Earth-sized planets in the habitable zone

2. Find out how many terrestrial-type and larger types of planets exist in their stars' habitable zones

3. Find planets (if any) in multiple-star systems

4. Determine the properties of stars with planets

5. Determine the properties of the planets it finds

6. Determine the sizes of the orbits of those planets

Kepler's Future

Planet detection takes years of repeated observations, surveying hundreds of thousands of stars. *Kepler* was launched on a three-and-a-half-year prime mission, just barely allowing it to find planets with orbits about a year long. In 2012, that mission was extended at least another four years. Although none of the planets it has found so far are exactly like Earth, the mission has identified thousands of "super Earths," including one in the habitable zone of its star, and "mini-Neptunes" around distant stars. Now *Kepler* can turn its attention to searching out what are called "true Sun-Earth analogs." Those planets should have orbits about a year long, and their stars should be similar in classification to the Sun. Already scientists are predicting that those analogs will be found in the very near future, and once they are found, the next step will be to study their atmospheres for signs of life.

Preliminary results show that every star *Kepler* is studying is expected to have at least one planet. If you extend that out to the rest of our galaxy, it turns out the Milky Way could contain billions of other worlds. *Kepler* may have already made a preliminary detection of one of those worlds, and confirmation awaits further observations by researchers using the *Spitzer Space Telescope* as well as other space-based and ground-based telescopes.

Kepler's Greatest Hits (So Far)

Here are a few of the most interesting discoveries *Kepler* has made:

1. Kepler 47: a binary star system with two planets
2. Kepler 36: one star with two planets orbiting closely together
3. Kepler 20: a planet 2.4 times the size of Earth in the Habitable Zone of a star like our Sun
4. Kepler 22: the first star found with a planet in its habitable zone
5. Kepler 16b: a world with a *Star Wars*–style double sunset because it orbits two stars
6. Kepler 10-b: the first planet known to be rocky like Earth, with a 1-day orbit, and an ocean of lava in one hemisphere

CHANDRA X-RAY OBSERVATORY

X-Rays in Space

The universe has hot spots, places such as the centers of galaxies or near very compact stellar objects like black holes and neutron stars—where temperatures can reach millions of degrees. These superheated regions give off radiation in the form of x-rays and tell astronomers that whatever is emitting them is very energetic. X-ray emissions are difficult to detect from Earth's surface because our planet's atmosphere absorbs the incoming radiation. So a space-based satellite is the best way to study such objects as x-ray binaries—systems that contain a neutron star or a black hole paired with a normal companion star, or active galactic nuclei (AGN) that emit copious amounts of x-rays.

The History of X-Ray Astronomy

X-ray astronomy began with research rockets in 1949, carrying instruments designed to study solar x-ray emissions. The science really took off in the 1960s as rockets and high-altitude balloons looked beyond the Sun at other sources of x-rays in space. In the 1970s, the *Uhuru* was launched as the first satellite specifically designed for x-ray studies. It was followed by *Ariel 5*, *SAS-3*, *OSO-8*, and *HEAO-1*. Since then, the field of x-ray astronomy has seen the launch of seven more probes, including *EXOSAT*, *ROSAT*, *ASCA*, *BeppoSAX*, and others. Today, astronomers use *XMM-Newton* from the European Space Agency and the *Rossi X-Ray Timing Explorer*, *SWIFT*, and *Chandra X-Ray Observatory* satellites launched by NASA.

The *Chandra X-Ray Observatory*

One of the best known of the current x-ray observatories is called *Chandra*, for the Indian astrophysicist Subramanyan Chandrasekhar (1910–1995). His name means "luminous" in Sanskrit. Chandrasekhar is best known for his work on the structures of stars and, in particular, of very evolved stars such as white dwarfs. He determined something called the *Chandrasekhar Limit*, which is the maximum mass that a white dwarf can possess before collapsing to become a neutron star or a black hole. The observatory was named for him in honor of his work and contributions to the understanding of evolved stars and was launched into orbit aboard the space shuttle *Columbia* in 1999 as one of NASA's Great Observatories. Since then, it has observed thousands of x-ray sources and surveyed some of the most distant sources of this radiation in the universe.

One of *Chandra*'s projects is a survey of some of the biggest black holes known to exist in the universe. These supermassive objects have long been known to exist at the hearts of galaxies. *Chandra*'s observations show that there may be a lot more of these behemoths than originally thought, and they could contain up to a billion times the mass of our Sun. These supermassive black holes lie in galaxies that are part of hugely massive clusters permeated with clouds of hot gas. The central black holes are very active, and their outbursts keep the gas superheated and so it cannot collapse to form new stars. It also means that the black holes themselves are gulping down the gas in order to stay active.

The *Chandra X-Ray Observatory* is one of the great astronomy workhorses of our time, along with the *Hubble Space Telescope* and *Spitzer Space Telescope*. Its contributions to high-energy astronomy have transformed our view of the universe.

How *Chandra* Works

Just as many other observatories, *Chandra* is outfitted with mirrors. But its mirrors are not flat because x-rays will simply slam right through them. Instead, *Chandra*'s engineers created long, barrel-shaped mirrors that allow x-rays to ricochet off the surfaces and onto detectors. These instruments record the direction and energy of the photons, which tells about conditions at the source of the x-rays. As of 2012, the observatory has been in orbit for thirteen years, and despite a few minor issues with its spacecraft insulation, *Chandra* is expected to perform for another decade.

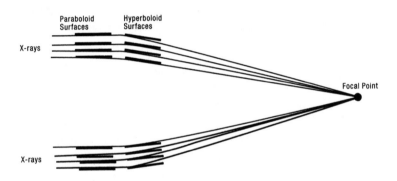

Chandra's mirrors are long, tubular-type surfaces that focus the incoming x-rays into the spacecraft's detectors for study.

Chandra's Targets

The *Chandra X-Ray Observatory* studies many different types of emission sources. In our own solar system, it cannot look at the Sun, but it can focus on x-rays from Earth's outer atmosphere as it orbits through this region high above our planet. They are caused by high-speed collisions between hydrogen atoms from our atmosphere and energized atoms (called ions) of carbon, oxygen, and neon streaming away from the Sun. Elsewhere in the solar system, *Chandra* can focus in on x-rays from comets as they traverse the solar wind, emissions caused by solar x-ray impacts on the Moon, and emissions from the collision of the solar wind particles with the gigantic, powerful magnetic fields of Jupiter and Saturn.

Beyond the solar system, x-rays stream from many places:

- Energetic stars
- Supernova explosions
- Rotating neutron stars that energize particles in their environment
- The centers of black holes
- Quasars
- Vast clouds of extremely hot gas that flow between galaxies in clusters

In addition, *Chandra* surveys the x-ray background. This is a faint wash of x-ray emissions from distant sources such as galaxies with active black holes in binary star systems, hot gases within galaxies, and supernova explosions. In the very distant universe, *Chandra* has detected x-rays from active galaxies and quasars so far away that the emissions have traveled for more than 12 billion years.

Chandra and Dark Matter

Dark matter is, as its name implies, dark. It can't be seen. This mysterious substance permeates the universe, and yet no one has been able to detect it directly. We can measure its gravitational effects on baryonic matter—the material we do see. Dark matter is highly concentrated in galaxy clusters, and when individual galaxies or clusters collide and merge, friction between the hot gas clouds in these objects will separate them from their dark matter. This separation can be observed by optical telescopes that search out the distortions of galaxy shapes by the gravitational lensing effect of the dark matter. *Chandra* detects the x-rays given off by the baryonic matter as it is superheated, giving astronomers a very good idea about where dark matter exists in those clusters.

SPITZER SPACE TELESCOPE

An Infrared Eye on the Sky

The universe is illuminated by emissions from distant objects. We see the stars, planets, nebulae, and galaxies in visible light. That's only a tiny part of all the light that streams from objects in the cosmos across a range called the electromagnetic spectrum. Some emissions, such as ultraviolet, x-ray, gamma ray, and infrared, are absorbed by our planet's atmosphere, making observations from the ground very difficult. This is particularly true for infrared light, also known as IR. Some IR makes it to the ground and can be detected using specialized instruments at high altitude observatories, but much of it can't be detected here. To detect infrared light astronomers send specialized observatories into space, such as the *Spitzer Space Telescope*, launched by NASA in 2003 and still active today.

Infrared Observatories

There are a number of infrared-enabled ground-based observatories in operation today, including several in Hawaii (Gemini, IRTF), Chile (VISTA, Gemini), and Wyoming (WIRO). Most other facilities have operated in space, including the European Space Agency's *Herschel Space Observatory*, which is sensitive to far-infrared and submillimeter (radio) emissions, and NASA's *Wide-Field Infrared Survey Explorer* (WISE), which operated from 2009 to 2011. The first IR-specific observatory in space was the *Infrared Astronomical Satellite* (IRAS), built by the United States, Great Britain, and the Netherlands. For ten months in 1983, it scanned most of the sky four times. Among other things, it was able to pierce through the clouds of gas and dust at our galaxy's heart to spot the center of the Milky Way.

The Science Behind *Spitzer*

Infrared light was first detected by Sir Frederick William Herschel in 1800. He was experimenting with filters that would let him look at sunspots when he tried to pass sunlight through a red filter. To his surprise, he detected heat. He called this radiation "calforic rays," and in time they were referred to as infrared. This light can pass through thick clouds of gas and dust, revealing warm objects inside.

There are three types of infrared that astronomy investigates:

1. *Near infrared*, which can be detected from ground-based observatories. It comes from anything in the universe that gives off any kind of heat and includes cool red stars and giants. A near-IR detector can see right through interstellar dust clouds.

2. *Mid-infrared*, which can be detected from a few very high-altitude observatories on Earth but is also studied from space. It indicates the presence of cooler objects such as planets, asteroids, comets, and disks around newborn stars. Clouds of dust warmed by nearby stars also show up in the mid-infrared.

3. *Far infrared* is emitted by objects such as thick, cold clouds of gas and dust that exist in the interstellar medium. Many are hiding proto-stars, which are just beginning to form. The action at the center of our galaxy also heats up these clouds, and the dust is obvious in far-infrared light. The plane of our galaxy (which is the region that contains the spiral arms and extensive gas and dust clouds) glows brightly in the far infrared, as well.

The *Spitzer Space Telescope* was built with sensitivity to near-, mid-, and some far-infrared light. Its primary mission ended when it ran out of liquid helium to keep the telescope's instruments at their coldest. However, *Spitzer* continues to gather data with instruments that do not need such high levels of cooling, and the Spitzer Warm Mission will continue.

Lyman Spitzer Jr.: the Man Behind the Telescope

Lyman Spitzer Jr. is best known for his championing of the *Hubble Space Telescope*. He was trained in astrophysics, taught at Yale and Princeton, and spent much of his research time studying the interstellar medium, which is not empty but filled with gas and dust. He also explored star-forming regions. He created the Princeton Plasma Physics Laboratory and worked on finding ways to use nuclear fusion for power generation. In 1946, Spitzer proposed that NASA consider building a space telescope that would be free of the blurring effects of our atmosphere. The agency ended up building four "Great Observatories," of which *Hubble* was one. The others were the *Compton Gamma-Ray Observatory* (CGRO), *Chandra X-Ray Observatory*, and *Space Infrared Telescope Facility* (SIRTF). Eventually, *SIRTF* was renamed in honor of Spitzer for his many contributions to astronomy and astrophysics.

Spitzer's Greatest Hits

The *Spitzer Space Telescope* has been wildly successful. In 2005 it was the first telescope to directly detect light from two exoplanets orbiting distant stars. The data indicated that these stars were so-called "hot Jupiters" with temperatures of 727°C (1,340°F). Elsewhere, *Spitzer* may have detected the heat of a collision between two distant planets circling a young star. This could mirror similar collisions that occurred early in the history of our own solar system. In one of its most fascinating observations, *Spitzer* captured light from what could be the oldest stars in the universe.

Peering Into a Star Birth Crèche

Star birth takes place hidden from our view behind thick clouds of gas and dust. Infrared-enabled telescopes such as *Spitzer* take advantage of infrared light's ability to pass through dust clouds to reveal the embryos of future stars. In many cases, these stellar seedlings are smothered in huge, pillar-shaped formations that are created when strong radiation from their older, more massive stellar siblings carves away and destroys the star birth clouds. Where optical telescopes would see only a pillar of darkness outlined in starlight, *Spitzer* lifts the dusty curtain hiding the process of star birth and reveals the stellar newborns.

Infrared Trivia

Human beings glow in infrared light. That's because we give off thermal radiation (heat), which can be picked up by an infrared detector. On Earth, doctors use infrared as a diagnostic tool, and most of us are aware of infrared-sensitive night-vision binoculars and cameras. NASA and other space agencies equip Earth-sensing satellites with IR detectors to monitor weather and atmospheric events, ocean temperatures, and other surface changes.

FERMI

Gamma-Ray Astronomy from Space

Some of the most exotic phenomena in the universe give off prodigious amounts of radiation in the form of gamma rays. These are the most energetic form of electromagnetic radiation and are emitted by cosmic objects with the most extreme densities and strongest magnetic and gravitational forces. All but the most energetic cosmic gamma rays are absorbed by Earth's atmosphere, so astronomers must launch satellites to study them from space.

There's a whole cosmic zoo of objects and events that produce gamma rays. Gamma-ray bursts (GRBs), seen about once a day, have as much energy as our Sun emits in its 10-billion-year lifetime. Some of these bursts seem to be associated with the deaths of massive stars—called supernova explosions. Supernovae are also sources of nuclear gamma rays—specific signatures of chemical elements that are created during the stars' explosive death throes. The stellar cores that collapse in these explosions form neutron stars—city-sized objects as dense as atomic nuclei—which can emit gamma rays in jets that are channeled by magnetic fields trillions of times as strong as those found on Earth. When these objects spin rapidly, they are known as *pulsars*.

The most prolific sources of gamma rays are supermassive black holes in the cores of distant galaxies, which emit powerful jets stretching out over tens of thousands of light-years. If Earth happens to be in the line of sight of this radiation, we see what is termed a *blazar*.

Searching Out Gamma-Ray Sources

Gamma-ray astronomy began in the 1960s with the *Explorer XI* satellite. It was followed by the *Orbiting Solar Observatory, Vela,* and dozens of other missions that had instruments sensitive to gamma rays. The first mission focused largely on this part of the electromagnetic spectrum was the *Compton Gamma-Ray Observatory*, which extensively surveyed the gamma-ray sky from 1991 to 2000. It was followed by the *BeppoSA* observatory, which was also sensitive to gamma-ray bursts. The European Space Agency launched the *International Gamma-Ray Astrophysics Laboratory (INTEGRAL)* in 2002, and in 2008, NASA launched the *Fermi Gamma-Ray Space Telescope. Fermi's* Large Area Telescope brings a wide-area sky view of gamma-ray emissions, while its Gamma-Ray Burst Monitor focuses on sudden outbursts.

Gamma Rays and You

Gamma radiation can be deadly. In space, if you were anywhere near a gamma-ray burst or supernova explosion, the stream of gamma rays would kill you instantly. However, the properties of this radiation that kill living cells can also be used to locate and destroy cancers in the human body. Some food producers also use gamma radiation to kill bacteria in fresh foods.

Fermi's Universe

The gamma-ray universe is a violent one. *Fermi*, which was named for Italian high-energy physicist Enrico Fermi (1901–1954), surveys the entire gamma-ray sky every three hours. The energetic universe it sees changes on short time scales, and so *Fermi* can detect sudden changes in gamma-ray intensity almost as soon as they begin. For example, one of its first major discoveries was a pulsar that seems to be emitting almost all of its energy in gamma rays and is spinning very rapidly—once every 316.86 milliseconds. This gamma-ray pulsar emits more than a thousand times the energy output of the Sun. *Fermi* has since discovered more than a hundred additional gamma-ray pulsars, some which spin around as fast as once every millisecond. High-energy gamma rays have also been detected from many supernova remnants. This indicates that the initial supernova blast wave rapidly accelerated charged particles to near the speed of light.

One of the most intriguing *Fermi* projects is to make a census of distant stars in the universe by observing blazars, extremely energetic ancient galaxies that appear very bright. Gamma rays from these blazars shine through a fog of cosmic background light that was emitted by stars that no longer exist. However their light continues to race across the universe creating a stellar fog. As gamma radiation travels through this so-called extragalactic background light (EBL), it disappears. Astronomers can track the dimming much like ship captains can watch the light from a lighthouse get swallowed up in a dense fog. By measuring the loss of gamma-ray intensity as it travels, it's possible to figure out the thickness of the starlight fog, which then tells how much starlight has ever shone in the universe.

The *Fermi Gamma-Ray Space Telescope* is nearly at the end of its original five-year planned mission. It should last at least another ten years, and astronomers are planning to use *Fermi* to make many more observations of distant and exotic gamma ray–emitting objects.

Gamma-Ray Bursts

The high-energy universe of gamma rays constantly flickers with outbursts from energetic objects. Gamma-ray bursts (GRBs) are the most powerful of these events, and they're associated with distant galaxies that lie billions of light-years away from us. GRBs are very short events—lasting from around 10 milliseconds up to several minutes. They fade, but their afterglows can be detected in most other wavelengths of light. The *Fermi* observatory is uniquely equipped to study these giant explosions. For example, it recorded a gamma-ray burst that occurred on September 16, 2008, one of the most powerful ever recorded. That event released more power and radiation than 9,000 supernova explosions, and its gas jets streamed away at just under the speed of light. This magnificent explosion occurred more than 12 billion light-years away and lasted for twenty-three minutes before starting to fade.

Gamma Rays on Earth

Fermi has also studied gamma-ray blasts detected near the locations of thunderstorms on Earth. These are called terrestrial gamma-ray flashes. The exact mechanism for these outbursts seems to come from strong static electricity bursts that accelerate electrons in the atmosphere; these electrons, in turn, collide with other atoms. This produces a cascade of gamma rays high above thunderstorms. Scientists are studying the effects of these emissions to find out if they pose any threat to aircraft crews and passengers flying nearby during intense thunderstorms.

Fermi and the Sun

The Sun can release incredible amounts of energy during outbursts called solar flares. On March 7, 2012, it unleashed a powerful X5.4-class solar flare. Not only did this flare produce huge amounts of x-rays, but astronomers using *Fermi* detected so much gamma radiation from this one flare that for a short time the Sun was the brightest gamma-ray object in the sky.

THE FUTURE OF ASTRONOMY

It's Looking Up

With all the major discoveries in astronomy that are announced each year, is it possible that "it's all been discovered"? No. Astronomers have barely scratched the surface of what there is to learn about the universe. There's a lot more exploring and explaining to do. Furthermore, there is so much data pouring in from our current crop of observatories that there's plenty of work for researchers, graduate students, advanced amateurs, and even some good undergraduate and high school students to do. There is so much information streaming in through our observatories that serendipitous discoveries await any astronomy data miners who dig into those observational treasure troves.

What's On the Drawing Boards?

The future of astronomy is taking shape now. Over the next decade, we will see and hear about discoveries made with the Atacama Large Millimeter Array (ALMA). Extended installations such as the Square Kilometer Array (SKA) and others are being planned and built in radio-quiet areas of our planet. In the slightly more distant future, astronomers plan to build radio telescope arrays on the far side of the Moon, well away from the radio frequency interference that pollutes much of Earth's airwaves and makes things difficult for specific types of radio astronomy.

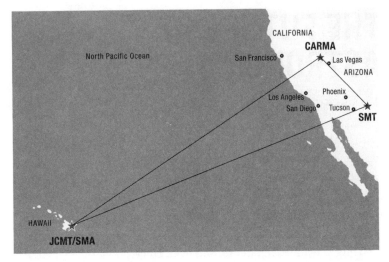

Very large baseline interferometry (VLBI) in radio astronomy links together several telescopes to focus in on very small regions of space with high precision. Telescope arrays are already linking many instruments together to produce the gathering power of a telescope the size of the array. Arrays will play a huge role in the future.

In optical astronomy, the next big thing on the ground is the Thirty-Meter Telescope (TMT) to be built on the Big Island of Hawaii. At the heart of this project is a thirty-meter (almost 100-foot) segmented mirror, which will be able to carry out optical and infrared astronomy, returning some of the sharpest images yet of the universe. The United States, China, and India are partners in the project, as well as the National Astronomical Observatory of Japan.

Infrared space astronomy will get a boost when the *James Webb Space Telescope* (*JWST*) goes into orbit sometime in the next decade. It's the successor to the *Hubble Space Telescope* and has a planned five-to-ten-year mission. Like the *Spitzer Space Telescope* and other

ASTRONOMY 101

infrared-enabled observatories, *JWST* will continue astronomy's look at the processes of star birth and star death, attempt direct imaging of extrasolar planets, and search out light from the earliest epochs of the universe.

Planetary Exploration

The exploration of our solar system will continue as NASA sends missions to Mars and continues the missions to Saturn and Pluto that are already underway. China is taking a long look at lunar exploration, and there are rumors that it is also interested in Mars and maybe a few asteroids. Its modernization includes a new concentration on astronomy and so China is building new observatories and educating masses of new researchers. Chile and Argentina are partnering on major new observatories in South America, and the European Space Agency is planning new missions to study Earth's climate and a Mars Sample Return mission as a follow-up to its highly successful *Mars Express* program. In addition, ESA is planning the ambitious Cosmic Vision roadmap of missions to search out dark matter, study the Sun, and visit Jupiter.

Human Spaceflight

For many decades, space exploration was limited to people who trained for the rigors of spaceflight and went into Earth orbit or to the Moon in order to perform specific tasks in those places. Today there are plans afoot to make near-Earth trips possible for "regular" people. Of course, they will be costly at first, but someday trips to orbital hotels and resorts on the Moon will become more commonplace. Trips to Mars are another big goal for the world's space agencies. Visionaries have come up with Mars exploration scenarios that include lengthy stays for geological exploration, and eventually the construction of permanent habitats and colonies on the Red Planet.

Distant Horizons

The exploration of deep space is still very much on everyone's agenda. ESA has on the drawing board an exoplanet search mission called Plato and a gravitational wave program (with NASA) called LISA. In the United States, the 8.4-meter telescope called the Large Synoptic Survey Telescope is in the planning stages. It will image the entire sky every few nights. It is planned for construction on Cerro Pachón in Chile, the same mountain that houses the Gemini South Telescope. Once built, it will map the Milky Way, measure the light from distant objects to detect weak gravitational lensing, look for near-earth asteroids and Kuiper Belt objects (out beyond the orbit of Neptune), and be on the lookout for short-lived events such as nova and supernova outbursts.

One thing that is certain is that astronomy of the future will not be limited to a few countries. Astronomy as a science has changed from "lone geniuses at the telescope" to large, multi-national collaborations of people who work together for many years toward a common scientific goal. The countries that participate in these Big Science projects will require talented, well-educated students to learn about astronomy and continue their work into the far future. In the United States, science, technology, engineering, and math (STEM learning) all aim to bring students into the future. All of these disciplines are needed to continue humanity's reach into the skies as we yearn for a better understanding of the cosmos.

Astronomy Education

The future of astronomy depends on students who are interested and seek out education in the basics of this science. Yet in the United States, astronomy topics are few and far between in most educational curriculum standards. Astronomy is well known as a gateway science leading to interest in the following disciplines:

- Physics
- Mathematics
- Computer programming
- Chemistry
- Geology
- Biology
- Engineering

Astronomers and outreach professionals at planetariums and science centers, as well as amateur astronomers, are working with such organizations as the Astronomical Society of the Pacific to provide healthy educational support for astronomy and its related STEM topics both in schools and among the general public.

YOU CAN DO ASTRONOMY

The Future's in the Stars

Doing astronomy isn't limited to professional observers and big telescopes. Hundreds of thousands of amateur astronomers around the world also watch the sky. Some simply go outside, look up, and enjoy the view. Others set up telescopes and do astrophotography, and a few even contribute to scientific research. It's easy to get started—you just step outside on a clear, dark night and look up. If you're a novice stargazer, what you see will amaze you, and before long you'll want to know more about the objects you observe.

What's Up There

During the day, there's usually only one star that's visible: the Sun. The Moon also appears in daylight during part of the month. At night, you have the planets, the Moon during part of the month, the stars, and galaxies to explore. Star charts can help you find your way around the sky. These are maps that show what's up in the sky in the month or season you're observing. There are many sources of star charts online as well as in such magazines as *Sky & Telescope* and *Astronomy*.

People often think that they have to buy huge telescopes and cameras to "do" astronomy. That's not so. Naked-eye observing is the best way to get started. This kind of astronomy is as easy as looking at the sky and learning your way around the stars. There are many good books for casual observers that help you explore the stars and constellations in more detail. One of the best and easiest to learn from is *Find the Constellations* by H. A. Rey. Originally written for children and their parents, it is a great beginner's book. *The Stars: A New Way to See Them*, also by Rey, has more details and is written for older children and adults.

Equipment Fever

If you want to get some equipment, think about what you want to see. What you're trying to do is make distant things look closer so you can see them better. The best way to start is with a pair of binoculars. They help magnify the view and are easy to use. Once you get to know the sky and have some favorite objects you want to study in more detail, *then* think about getting a telescope.

The type of telescope you buy will depend on which celestial objects you want to view. Before you plunk down hard-earned money on optics, look through other people's telescopes. Ask them a lot of questions, and do your homework before you buy. There are good equipment-buying tutorials from *Sky & Telescope* and *Astronomy* magazines, so check those out, too.

What to Observe

The easiest things to observe in the sky are the planets, particularly Mercury, Venus, Mars, Jupiter, and Saturn. They're visible to the naked eye, and they stand out because they're usually brighter than the stars. If you want to try finding Uranus and Neptune, you'll need a good backyard-type telescope. The next things to spot are double stars. After that, you can look for star clusters such as the Beehive in Cancer, or the Double Cluster in Perseus. If your skies are really dark, you should be able to make out the Milky Way, particularly in the summer and wintertime skies. Those are just a few of the many gorgeous sights that await you as you explore the skies!

Easy Targets to Observe

If you have binoculars or a small telescope, check out these sky sights:

1. The Moon—you can see individual craters and other surface features
2. Mars—through a small telescope, you can spot its dark and bright areas and its polar caps
3. Jupiter—through binoculars you can look for its four brightest moons
4. Saturn—look for its rings through binoculars and the bright moon Titan through a telescope
5. The Andromeda Galaxy—near the W-shaped constellation Cassiopeia. It's most visible from August through part of March; it is just barely visible to the naked eye
6. The Orion Nebula—best viewed from November through mid-April, just below the three stars in Orion's belt
7. Alcor and Mizar—a double star in the bend of the handle of the Big Dipper
8. Albireo—a beautiful double star in the constellation Cygnus, best visible at night July–November

Observing Mercury from Earth

It's very easy to see five of the planets from Earth with the naked eye. They look like bright points of light against the backdrop of stars (Uranus and Neptune are too faint to see without a good telescope). Mercury is challenging because it orbits close to the Sun. However, there are certain times when observers can see it for a few days. Look for Mercury at sunset during a time called *greatest eastern elongation*, and at sunrise during *greatest western elongation*. Elongation is a term that refers to the points in Mercury's orbit in the sky as seen from Earth when the planet is most distant from the Sun. The table here gives you the best time, date and position (in degrees from the horizon), and brightness of Mercury over the next few years. So, on October 9, 2013, for example, it will appear as a bright pinpoint of light in the western sky 25.3 degrees *east* of the Sun (which is setting). In November, it will appear in the morning sky on November 18, 19.5 degrees *west* of the Sun (so look for it before sunrise).

Best Times to Observe Mercury

TIME	DATE	ELONGATION	MAGNITUDE
Evening (after sunset)	2013 Oct 9	25.3°E	+0.2
Morning (before sunrise)	2013 Nov 18	19.5°W	-0.3
Evening (after sunset)	2014 Jan 31	18.4°E	-0.3
Morning (before sunrise)	2014 Mar 14	27.6°W	+0.4
Evening (after sunset)	2014 May 25	22.7°E	+0.7
Morning (before sunrise)	2014 Jul 12	20.9°W	+0.6
Evening (after sunset)	2014 Sep 21	26.4°E	+0.3
Morning (before sunrise)	2014 Nov 1	18.7°W	-0.3
Evening (after sunset)	2015 Jan 14	18.9°E	-0.4
Morning (before sunrise)	2015 Feb 24	26.7°W	+0.3
Evening (after sunset)	2015 May 7	21.2°E	+0.5
Morning (before sunrise)	2015 Jun 24	22.5°W	+0.7
Evening (after sunset)	2015 Sep 4	27.1°E	+0.5
Morning (before sunrise)	2015 Oct 16	18.1°W	-0.3
Evening (after sunset)	2015 Dec 29	19.7°E	-0.3

One note of caution: Make sure the Sun is not in the sky when you look for Mercury. You don't want to injure your eyes by looking directly at the Sun.

Light Pollution: The Astronomer's Bane

Every night people pollute the sky with unnecessary lights. This washes out the view of the sky and is called light pollution. It's not just a threat to seeing the stars—it also causes health problems and costs money to keep unnecessary lights on where they're not needed. Light pollution lets people living in urban areas see only the brightest stars and planets in the sky. In many large cities, people only see a few stars, and there are many who have never seen the Milky Way.

No one is advocating that we abolish the light bulb. There are ways to safely light our homes, streets, and landscapes and keep our view of the stars. People can help by turning off unnecessary lights around their homes. If lights are necessary, aim them directly where they are needed. And remember: The stars are everyone's legacy.

Observing Target: Eclipses

No words can completely describe the experience of seeing an eclipse. People travel around the world to experience solar eclipses, particularly if the eclipse is total. There are many eclipse chasers who travel the world to watch as the Moon's disk slowly pushes across the Sun's face so that they can experience a few minutes of darkness during the day.

There are two kinds of eclipses: lunar and solar. A lunar eclipse occurs when the Moon moves through Earth's shadow. It takes several hours, and during that time the Moon appears to turn a dark, rusty, red color. A solar eclipse occurs when the Moon moves between Earth and the Sun. If you're observing from the path of the shadow on Earth's surface during a total solar eclipse, the Moon blots out everything except the solar corona—the Sun's thin upper atmosphere. The temperature at your observing site drops, and for a short time, you can see the brightest stars and planets in the deep twilight sky. There are also events called annular solar eclipses, where the Moon doesn't quite cover the entire Sun. Instead, observers see a "ring of fire" during totality. To find out when the next eclipse of any kind occurs in your area and for complete information on these fantastic events, visit *www.mreclipse.com*.

BIBLIOGRAPHY/REFERENCES

Print references used throughout the book:

Beatty, J. K., C. C. Petersen, and A. C. Chaikin, *The New Solar System*, Fourth Ed., Cambridge University Press/Sky Publishing, 1998.

Bennett, Jeffrey O., Nicholas Schneider, and Mark Volt, *Cosmic Perspective*, Addison Wesley, 2004.

Brown, M. *How I Killed Pluto and Why It Had It Coming*, Spiegel and Grau, 2010.

Hartmann, W. K., *A Traveler's Guide to Mars*, Workman Press, 2003.

Lankford, J., ed., *History of Astronomy*, Cambridge University Press, 1997.

Pasachoff, J., *A Field Guide to Stars and Planets*, Fourth Ed., Houghton Mifflin, 2006.

Petersen, C., and J. Brandt, *Hubble Vision*, Second Ed., Cambridge University Press, 1998.

Petersen, C., and J. Brandt, *Visions of the Cosmos*, Cambridge University Press, 2003.

Rees, M., *Universe: The Definitive Visual Guide*, Smithsonian Press/DK, 2012.

Sagan, C. *Pale Blue Dot: A Vision of the Human Future in Space*, Random House, 1994.

Rey, H. A. *Find the Constellations*, Houghton Mifflin, 2008.

Rey, H. A. *The Stars: A New Way to See Them*, Houghton Mifflin, 2008.

Strom, R. G., and A. L. Sprague, *Exploring Mercury*, Springer/Praxis, 2003.

Recommended Periodicals

Astronomy magazine (*www.astronomy.com*)

Sky & Telescope magazine (*www.skyandtelescope.com*)

Online References

THE SOLAR SYSTEM

www.solarsystem.nasa.gov/planets/index.cfm
www.starchild.gsfc.nasa.gov/docs/StarChild/solar_system_level1/
solar_system.html
www.universetoday.com/15959/interesting-facts-about-the-solar-system

THE SUN

www.nasa.gov/mission_pages/sunearth/index.html
www.solarsystem.nasa.gov/planets/profile.cfm?Object=Sun
www.sohowww.nascom.nasa.gov
www.sdo.gsfc.nasa.gov
www.nasa.gov/mission_pages/stereo/main/index.html

SPACE WEATHER

www.nasa.gov/mission_pages/sunearth/index.html
www.spaceweather.com
www.haystack.mit.edu/atm/index.html

MERCURY

www.solarsystem.nasa.gov/planets/profile.cfm?Object=Mercury
www.nasa.gov/mission_pages/messenger/main/index.html

VENUS

www.solarsystem.nasa.gov/planets/profile.cfm?Object=Venus

EARTH

www.earthobservatory.nasa.gov
www.nasa.gov/topics/earth/index.html

THE MOON

www.solarsystem.nasa.gov/planets/profile.cfm?Object=Moon
www.nasa.gov/topics/moonmars

MARS

www.mars.jpl.nasa.gov
www.nasa.gov/mission_pages/mars/main/index.html
www.solarsystem.nasa.gov/planets/profile.cfm?Object=Mars

JUPITER

www.solarsystem.nasa.gov/planets/profile.cfm?Object=Jupiter
www.solarsystem.nasa.gov/galileo
www.voyager.jpl.nasa.gov

SATURN

www.solarsystem.nasa.gov/planets/profile.cfm?Object=Saturn
www.saturn.jpl.nasa.gov
www.voyager.jpl.nasa.gov

URANUS

www.solarsystem.nasa.gov/planets/profile.cfm?Object=Uranus
www.voyager.jpl.nasa.gov

NEPTUNE

www.voyager.jpl.nasa.gov
www.nineplanets.org/neptune.html
www.hubblesite.org

PLUTO

www.nasa.gov/mission_pages/newhorizons/main/index.html
www.solarsystem.nasa.gov/planets/profile.cfm?Object=Pluto

COMETS

www.solarsystem.nasa.gov/planets/profile.cfm?Object=Comets

METEORS AND METEORITES

www.skyandtelescope.com/observing/objects/meteors
www.stardate.org/nightsky/meteors
www.nineplanets.org/meteorites.html

ASTEROIDS

www.neo.jpl.nasa.gov
www.iau.org/public/nea
www.solarsystem.nasa.gov/planets/index.cfm

THE STARS

www.imagine.gsfc.nasa.gov/docs/science/know_l2/stars.html
www.stars.astro.illinois.edu/sow/sowlist.html

STAR CLUSTERS

www.hubblesite.org/explore_astronomy

STAR BIRTH

www.hubblesite.org/hubble_discoveries/hstexhibit/stars/starbirth.shtml
www.science.nasa.gov/astrophysics/focus-areas/
   how-do-stars-form-and-evolve
www.coolcosmos.ipac.caltech.edu

STAR DEATH

www.imagine.gsfc.nasa.gov/docs/science/know_l2/supernovae.html
www.burro.astr.cwru.edu/stu/stars_lifedeath.html

BLACK HOLES

www.hubblesite.org/explore_astronomy/black_holes
www.science.nasa.gov/astrophysics/focus-areas/black-holes/
www.damtp.cam.ac.uk/research/gr/public/bh_intro.html

GALAXIES
www.science.nasa.gov/astrophysics/focus-areas/what-are-galaxies
www.nasa.gov/mission_pages/GLAST/science/milky_way_galaxy.html
www.damtp.cam.ac.uk/research/gr/public/gal_home.html

GALAXY FORMATION
www.jwst.nasa.gov/galaxies.html
www.astr.ua.edu/keel/galaxies/galform.html

THE MILKY WAY
www.curious.astro.cornell.edu/milkyway.php
www.science.nasa.gov/astrophysics/focus-areas/what-are-galaxies
www.nasa.gov/mission_pages/GLAST/science/milky_way_galaxy.html

ACTIVE GALAXIES AND QUASARS
www.imagine.gsfc.nasa.gov/docs/science/know_l1/active_galaxies.html
www.stsci.edu/~marel/abstracts/abs_L2.html
www.bdaugherty.tripod.com/gcseAstronomy/active.html

DARK MATTER
www.science.nasa.gov/astrophysics/focus-areas/what-is-dark-energy
www.chandra.harvard.edu/xray_astro/dark_matter

GRAVITATIONAL LENSES
www.imagine.gsfc.nasa.gov/docs/features/news/grav_lens.html

THE BIG BANG
*www.science.nasa.gov/astrophysics/focus-areas/
  what-powered-the-big-bang*
www.exploratorium.edu/origins/cern/ideas/bang.html

A SCIENCE-FICTION UNIVERSE

www.nasa.gov/centers/glenn/technology/warp/warpstat_prt.htm
www.nasa.gov/centers/glenn/technology/warp/ideachev.html

EXTRATERRESTRIAL LIFE

www.seti.org

HISTORY OF ASTRONOMY

www.historyofastronomy.org

NICOLAUS COPERNICUS

www.scienceworld.wolfram.com/biography/Copernicus.html
www.plato.stanford.edu/entries/copernicus

GALILEO GALILEI

www.galileo.rice.edu
www.plato.stanford.edu/entries/galileo

JOHANNES KEPLER

www.kepler.nasa.gov/Mission/JohannesKepler
www.galileo.rice.edu/sci/kepler.html
www.galileoandeinstein.physics.virginia.edu/1995/lectures/kepler.html

THE HERSCHELS

www.williamherschel.org.uk

ISAAC NEWTON

www.galileoandeinstein.physics.virginia.edu/lectures/newton.html
www.newton.ac.uk/newtlife.html

HENRIETTA SWAN LEAVITT

www.womanastronomer.com/hleavitt.htm
www.web.mit.edu/invent/iow/leavitt.html
www.pbs.org/wgbh/aso/databank/entries/baleav.html

EDWIN P. HUBBLE

www.edwinhubble.com
www.quest.arc.nasa.gov/hst/about/edwin.html
www.hubblesite.org/reference_desk/faq/answer.php.id=46&cat=hst

ALBERT EINSTEIN

www.nobelprize.org/nobel_prizes/physics/laureates/1921/einstein-bio.html
www.einstein.biz
www.alberteinstein.info

JOCELYN BELL BURNELL

www.aip.org/history/ohilist/31792.html
www.starchild.gsfc.nasa.gov/docs/StarChild/whos_who_level2/bell.html

VERA COOPER RUBIN

www.aip.org/history/ohilist/5920_1.html
www.innovators.vassar.edu/innovator.html?id=68
www.phys-astro.sonoma.edu/brucemedalists/rubin/index.html

CLYDE TOMBAUGH

www.icstars.com/HTML/icstars/graphics/clyde.htm
*www.starchild.gsfc.nasa.gov/docs/StarChild/whos_who_level2/tombaugh
.html*

MIKE BROWN

www.gps.caltech.edu/~mbrown

ASTROPHYSICS AND ASTRONOMY

www.science.nasa.gov/astrophysics

www.imagine.gsfc.nasa.gov/docs/ask_astro/ask_an_astronomer.html
www.nrao.edu/index.php

ASTROBIOLOGY
www.astrobiology.nasa.gov

PLANETARY SCIENCE
www.science.nasa.gov/planetary-sciencewww.science.jpl.nasa.gov/
    PlanetaryScience/index.cfm
www.lpi.usra.edu/library/website.shtml

COSMIC TIME MACHINES
www.ifa.hawaii.edu/mko
www.obs.carnegiescience.edu
www.gemini.edu
tdc-www.harvard.edu/mthopkins/obstours.html

THE *HUBBLE SPACE TELESCOPE*
www.hubblesite.org

THE *KEPLER* MISSION
www.kepler.nasa.gov
www.thekeplermission.com

CHANDRA X-RAY OBSERVATORY
www.nobelprize.org/nobel_prizes/physics/laureates/1983/chandrasekhar-
    autobio.html
www.chandra.harvard.edu

SPITZER SPACE TELESCOPE
www.spitzer.caltech.edu
www.hubblesite.org/the_telescope/hubble_essentials/lyman_spitzer.php

THE *FERMI* MISSION

www.fermi.gsfc.nasa.gov/
www.nasa.gov/mission_pages/GLAST/main/index.html

THE FUTURE OF ASTRONOMY

www.tmt.org
www.nrao.edu/index.php
www.skatelescope.org

YOU CAN DO ASTRONOMY

www.astronomy.starrynight.com
www.bisque.com
www.darksky.org
www.shatters.net/celestia/
www.space.com
www.star-map.fr
www.stellarium.org

INDEX